Data-Driven Analytics for the Geological Storage of CO₂

Data-Driven Analytics for the Geological Storage of CO$_2$

Shahab D. Mohaghegh

CRC Press
Taylor & Francis Group
Boca Raton London New York

CRC Press is an imprint of the
Taylor & Francis Group, an **informa** business

CRC Press
Taylor & Francis Group
6000 Broken Sound Parkway NW, Suite 300
Boca Raton, FL 33487-2742

First issued in paperback 2020

© 2018 by Taylor & Francis Group, LLC
CRC Press is an imprint of Taylor & Francis Group, an Informa business

No claim to original U.S. Government works

ISBN 13: 978-0-367-73438-1 (pbk)
ISBN 13: 978-1-138-19714-5 (hbk)

Visit the Taylor & Francis Web site at
http://www.taylorandfrancis.com

and the CRC Press Web site at
http://www.crcpress.com

I would like to dedicate this book to Mr. Fareed Al Hashmi. As a top executive in the industry, Fareed was one of the first individuals that demonstrated the vision and the insight to recognize the value and the potential contribution of Artificial Intelligence and Machine Learning to reservoir engineering. He played a key role in supporting the efforts that have resulted in some of the most groundbreaking AI-related technologies in the oil and gas industry.

No other individual in the executive management level in our industry can claim to have contributed as much as Mr. Al Hashmi to the foundation of the application of Artificial Intelligence and Machine Learning in reservoir engineering, reservoir modeling and reservoir management.

I am proud to call him a friend and value all that I have learned from interacting with him.

Contents

Nomenclature

a	Tortuosity factor (1, default value)
c	Compressibility factor (1/Pa)
C	Land's coefficient (fraction, dimensionless)
f_{ia}	Fugacity of component i in the aqueous phase (MPa)
f_{ig}	Fugacity of component i in the gas phase (MPa)
h	Formation thickness (m)
H	Henry's constant (atm./mole fraction)
K	Permeability (mD, m²)
K_h	Horizontal permeability (mD, m²)
K_v	Vertical permeability (mD, m²)
K_{rg}	Gas relative permeability
m	Cementation factor
n	Saturation exponent
P	Pressure (Pa.)
P^0	Reference pressure (Pa.)
R_w	Resistivity of the formation water (ohm.m)
R_t	True formation resistivity (ohm.m)
S	Saturation (fraction, dimensionless)
S_{gi}^*	Gas saturation on the drainage curve (fraction, dimensionless)
S_{gt}^*	Trapped gas saturation (fraction, dimensionless)
$S_{g,max}$	Maximum gas saturation (fraction, dimensionless)
$S_{gt,max}$	Maximum trapped gas saturation (fraction, dimensionless)
S_{rg}	Residual gas saturation (fraction, dimensionless)
V	Bulk volume (reservoir m³)
V_{CO_2}	Theoretical maximum storage capacity (reservoir m³ of CO_2)
ϕ	Rock porosity (dimensionless)
ρ_{br}	Brine density @ reference pressure (kg/m³)
ρ^o_{br}	Brine density (kg/m³)

Acknowledgments

I would like to acknowledge and thank the contribution of many people. First and foremost, I like to thank Dr. Turgay Ertekin. Turgay initiated all my understanding of reservoir engineering and reservoir modeling and has been a mentor to me, ever since I met him in December of 1987. I will never be able to thank him enough.

I would like to acknowledge the contribution of many of my brilliant students. They have contributed significantly to all the academic work that I was involved in for many years. I am proud to have had many of them in my team. I specifically like to acknowledge the work of Dr. Alireza Haghighat, Dr. Vida Gholami, Dr. Alireza Shahkarami, Dr. Qin He, Dr. Yasaman Khazaeni, Dr. Shohreh Amini, and Daniel and David Moreno for their hard work and contributions to the work that is presented in this book.

I like to acknowledge and thank my wife, Narges Gaskari and my daughter Dorna Mohaghegh. I could not imagine my life without their support and patience. And finally, I would like to acknowledge and thank Professor Sam Ameri for his continuing support as the Chair of the Petroleum and Natural Gas Engineering at West Virginia University.

Author

Shahab D. Mohaghegh, a pioneer in the application of artificial intelligence and data mining in the exploration and production industry, is the president and CEO of Intelligent Solutions, Inc. (ISI) and professor of petroleum and natural gas engineering at West Virginia University. He holds BS, MS, and PhD degrees in petroleum and natural gas engineering.

He has authored three books and more than 170 technical papers and carried out more than 60 projects for National Oil Company (NOCs) and International Oil Company (IOCs). He is an SPE Distinguished Lecturer and has been featured in the Distinguished Author Series of SPE's *Journal of Petroleum Technology* (JPT) four times. He is the founder of Petroleum Data-Driven Analytics, SPE's Technical Section dedicated to data mining. He has been honored by the U.S. Secretary of Energy for his technical contribution in the aftermath of the Deepwater Horizon (Macondo) incident in the Gulf of Mexico and was a member of the U.S. Secretary of Energy's Technical Advisory Committee on Unconventional Resources (2008–2014). He represented (2014–2016) the United States in the International Standard Organization (ISO) on Carbon Capture and Storage.

Contributors

Shohreh Amini
Halliburton
Houston, Texas

Vida Gholami
Advanced Resources International Inc.
Houston, Texas

Alireza Haghighat
Eclipse Resources Corporation
State College, Pennsylvania

Amirmasoud Kalantari-Dahaghi
Department of Chemical and
   Petroleum Engineering
The University of Kansas
Lawrence, Kansas

Alireza Shahkarami
Department of Petroleum
   Engineering
Saint Francis University
Loretto, Pennsylvania

Introduction

Studying, analyzing, modeling, evaluating, and finally designing the geological storage of carbon dioxide (CO_2) requires a fundamental and robust understanding of the flow of fluids in the crust of the Earth, also known as fluid flow through porous media. This is a topic that has been the centerpiece of the exploration and production (upstream oil and gas) industry for the past several decades. No other discipline can be compared to reservoir engineering and reservoir modeling when it comes to fluid flow through porous media, especially when the interest is in formations that are thousands of feet below the water table, which is the distinguishing feature of reservoir engineering with hydrology. Therefore, the technology that has been developed and field-tested for more than fifty years (reservoir engineering and reservoir modeling) can contribute immensely to the storage of CO_2 in geological formations such as depleted oil and gas reservoirs, saline aquifers, and, most recently, unconventional plays such as shale.

The physics associated with the geological storage of CO_2 can generally be summarized with the same equations that have been the essence of hydrocarbon reservoir engineering and modeling for decades. This includes the diffusivity equation (a parabolic partial differential equation) along with some other equations to account for changes in temperature, chemical reactions, and any other interaction between the CO_2 and other fluids present in the rock or the rock itself. Therefore, how these equations have been dealt with in the oil and gas industry for the past several decades to model the phenomena of movement of fluids in rock under extensive pressure can be of tremendous value to those involved with the analysis, modeling, evaluation, design, and optimization of geological storage of CO_2.

It should be mentioned that a large number of studies, mainly conducted by non-reservoir engineering entities, seem to have the essence of "re-inventing the wheel." The authors of many of these studies end up writing their own codes and simulation models that at best can be described as academic exercises with a limited contribution to the industrial application of geological storage of CO_2. Of course, the main contribution of these studies should not be undermined, since they have contributed immensely to awareness about the importance of this technology as well as looking at the problem from a (sometimes) completely different point of view that may generate interesting results. However, the fact remains that their contribution to the real, industrial application of this technology is limited at best. The main reason behind this is the common academic practice of simplifying (and sometime grossly approximating) the problem in order to develop an understanding of the process. Some (not all) of the projects that are commonly funded by the

U.S. Department of Energy, and some (not all) of those that are conducted in some of the national laboratories, are good examples of such cases.

During the decades of building, history matching, evaluating, predicting, and planning injection and production operations in the upstream oil and gas industry, numerical reservoir simulation models were used extensively to accommodate these processes. Numerical reservoir simulation models are complex and elaborate computer models with the capability to build a digital rock (formation) under thousands of psi (atmosphere) of pressure containing multiple phases of fluid. The simulators are capable of modeling the drilling of injection and production wells as well as the injection and production of fluid from the formation under a large variety of conditions. Commercial versions of these computer simulators have gone through decades of test and examination by the industry, and have proven to be much more robust and accurate than in-house models made in academic environments.

These commercial models also provide user-friendly interfaces (some of them, not all of them) to be used by geoscientists, and present a large variety of capabilities that have been an essential part of operations by the oil and gas industry. Therefore, their use and applicability in the geological storage of CO_2 not only provides technical accuracy and reliability but also ease of use. However, it must be noted that a shortcoming of these commercial numerical simulation models is their lack of flexibility. If a given physical, chemical, or biological process has not already been formulated in these simulators, the user will not have the luxury of modifying the source code to incorporate the desired process. Nevertheless, situations that might require such a modification, especially in an industrial application, may prove to be rare.

Furthermore, there are issues associated with numerical reservoir simulation models (and any other models that might be used for the analysis, modeling, design, and optimization of the geological storage of CO_2) that need to be addressed. These issues can be summarized as follows:

- The uncertainty of reservoir characterization, which increases many fold in the case of saline aquifers, due to lack of existing wells;
- The accuracy of models that do not use large amounts of data;
- The requirement of uncertainty quantification;
- The extensive computational footprint of numerical simulation models, especially when they are elaborate enough to capture the physics.

Almost all of the above issues have been real issues in the upstream oil and gas industry for decades. The only exception may be the first item (saline aquifers), which can be substituted with exploration and modeling of green fields in the upstream oil and gas industry. How are these issues addressed in the oil and gas industry? And how can they be addressed for the geological storage of CO_2.

In this book the tools and techniques that are traditionally used in the petroleum industry to address these issues will be covered, along with the author's views on the shortcomings of these traditional tools as well as the state-of-the-art technologies that can markedly enhance the capabilities of these traditional tools. The main objective of this book is to demonstrate the use and application of data-driven analytics, specifically artificial intelligence and machine learning, and how they are applied to the geological storage of CO_2.

In the upcoming chapters we cover issues associated with the storage of CO_2 in depleted gas reservoirs, including an unconventional shale asset, saline aquifers, CO_2-EOR, and finally the issue of making sure that the CO_2 stays underground (monitoring) and how artificial intelligence and machine learning can augment the traditional techniques to help us address and successfully resolve all these issues.

1

Storage of CO$_2$ in Geological Formations

Shahab D. Mohaghegh, Alireza Haghighat, and Shohreh Amini

CONTENTS

Carbon dioxide (CO$_2$) is a colorless and odorless naturally occurring chemical compound that is vital to life on Earth. Naturally occurring CO$_2$ originates from processes such as decomposition, ocean release, and respiration. Anthropogenic or manmade CO$_2$ results from cement production, deforestation, and burning of fossil fuels such as coal, natural gas, and oil. While the production and consumption of naturally occurring CO$_2$ seem to be close to balance, the introduction of anthropogenic CO$_2$ to the atmosphere is altering the overall global balance of CO$_2$. After decades of scientific research, climate scientists have concluded that the abundance of CO$_2$ in the atmosphere is one of the main causes of potential climate change. Therefore, mitigating the production of anthropogenic CO$_2$ has become the focus of the scientific community.

As an indicator for global climate change, climate scientists refer to the global average temperature increase of approximately 0.78°C (1.4°F) in the twentieth century. Global temperature changes have been almost proportional to the change in CO$_2$ concentration in the atmosphere, which increased from 280 ppm in 1880 to 385 ppm in 2010 (1). This can be clearly seen in Figure 1.1. To mitigate the impact of anthropogenic CO$_2$, different solutions have been proposed, including an overall reduction in energy consumption, substitution of high-carbon fuels such as coal with low-carbon fuels such as natural gas, and finally capture and storage of CO$_2$ in geological formations. It seems obvious that no single solution will be the answer, and multiple solutions must be incorporated, simultaneously.

Despite all efforts to shift energy sources to renewable and atmosphere-friendly alternatives, fossil fuels are still the most essential source of energy for industries and transportation. Considering the growth in demand, fossil fuel consumption will continue to increase and, as a result, concerns about greenhouse gas emissions and their impact on global warming and climate change are increasing.

Carbon capture and storage (CCS) may be considered a transitional technology while fossil fuels are slowly replaced by more environmentally friendly fuels.

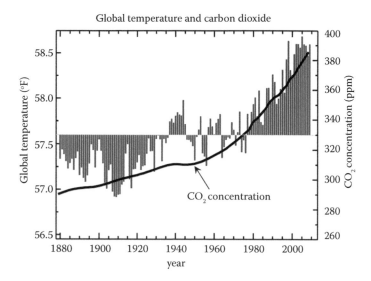

FIGURE 1.1
Global temperature and CO_2 concentration history (SI conversion for temperature: °C = (°F − 32) × 5/9). (National Climate Data Center. 2017. http://www.ncdc.noaa.gov/indicators/)

Given the fact that the transition cannot take place overnight and will require a considerable amount of time and effort to be expended, CCS could play an important role in mitigating the anthropogenic CO_2 issues that are faced today. The processes associated with CCS comprise separating CO_2 at the industrial level from power plants, refineries, cement plants and steel mills, transporting it to target storage locations, and finally injecting it into underground formations.

A viable means of reducing CO_2 in the atmosphere is to capture and concentrate CO_2 from large point sources such as power plants and petroleum refineries, and then store it by underground injection. This process is called geological carbon sequestration (GCS).

This book is dedicated to the application of solutions provided by big data analytics and data science to the geological storage of CO_2. As such, some preliminary topics, such as the basics of the geological storage of CO_2 as well as some of the solutions that are associated with big data analytics and data science, are covered to provide the necessary background for the main focus of the book, which is the application of big data analytics solutions to numerical modeling of the geological storage of CO_2.

1.1 Carbon Capture and Storage: CCS

There are several options for locations to store CO_2 in an underground geological structure: depleted oil and gas reservoirs, saline aquifers, and un-mineable coal-beds. Most recently, using shale plays as potential sites for carbon storage has been considered, and multiple studies have been conducted to evaluate their viability. Each of these geological structures has its own characteristics which must be extensively studied before any CO_2 is injected underground and stored for a long period (2).

Injected CO_2 can be stored through a number of different trapping mechanisms (3):

- Physical trapping, in structural or stratigraphic traps, where the free-phase CO_2 is physically trapped by the geometric structure of the reservoir and its cap-rock. This type of trapping is similar to hydrocarbon accumulations in a reservoir.
- Residual trapping, where the CO_2 is trapped in pore spaces due to capillary pressure forces.
- Solubility trapping, which takes place when injected CO_2 dissolves in the formation water.
- Mineral trapping, through which CO_2 precipitates as new carbonate minerals and therefore becomes immobile.
- Adsorption trapping, which happens when CO_2 is injected into coal-beds, where the CO_2 adsorbs onto the surface of the coal.

Although carbon sequestration seems to be a sensible way of reducing the ever increasing level of CO_2 in the atmosphere, the risk involved in this process is always a matter of concern. Safe sequestration is achieved when it is ensured that once the CO_2 is injected underground it will remain safely in the structure over a long geological period (thousands of years). In general, the risk involved in CO_2 sequestration in geological formations decreases as time passes. The risk associated with any geological sequestration is directly related to the geological uncertainties of the reservoir structure and operational practices, and therefore these items must be comprehensively studied for any CO_2 sequestration plan. There are several different potential sites for geological CO_2 storage, including depleted oil and gas reservoirs, deep saline aquifers, deep un-mineable coal seams and other unconventional plays such as shale, and finally, storage in association with CO_2-EOR (enhanced oil recovery) processes.

The number of commercial CCS projects that are currently operational is not large due to a lack of business and economic justification. This does not include CO_2-EOR projects, because CO_2-EOR projects include the production of petroleum and many companies throughout the world are engaged in such projects. Given the fact that the stored CO_2 lacks any commercial value, commencement of CCS projects for companies does not make sense from a profit point of view. Assembling new legal and regulatory frameworks could provide the required commercial justification for CCS deployment.

CCS projects consist of four different transitional phases (4):

1. Site selection
2. Operation
3. Closure and
4. Post-closure

"Site selection and development," the first phase, covers geological, commercial, and regulatory evaluation, which takes from approximately 3 to 10 years to purchase and secure space for surface facilities and geological storage. In addition, permission acquirement and infrastructure construction are completed in this phase. The "Operation" phase follows and includes CO$_2$ injection and further technical site monitoring. Depending on the storage capacity and operational designs, this phase may take decades. After "Operation," the "Closure" phase begins with implementation of different monitoring systems to assure no risk is associated with the stored CO$_2$. During "Closure," injection wells should be plugged following the removal of unnecessary infrastructures. Finally, the "Post-closure" phase is conducted with no involvement with the operator; some occasional observational or monitoring activities may be applied in this phase (4). This process is shown in Figure 1.2.

To begin the site selection phase, three fundamental characteristics of CO$_2$ storage sites are reviewed. The "storage capacity" indicates how much pore volume would be available in the reservoir for storage purposes. In order for the CO$_2$ to be represented in the supercritical phase, the required depth of the target formations is more than 800–1,000 m. Generally, storage capacity can be determined through original oil/gas in place (OOIP) calculations by knowing the area of the site, the formation thickness, the rock porosity, the density of the CO$_2$ under the storage conditions, storage efficiency (maximum CO$_2$ saturation), and the rock/brine compressibility factor.

Depending on whether the formation is composed of carbonate or clastic rock, favorable minimum porosity values for CO$_2$ storage sites vary between

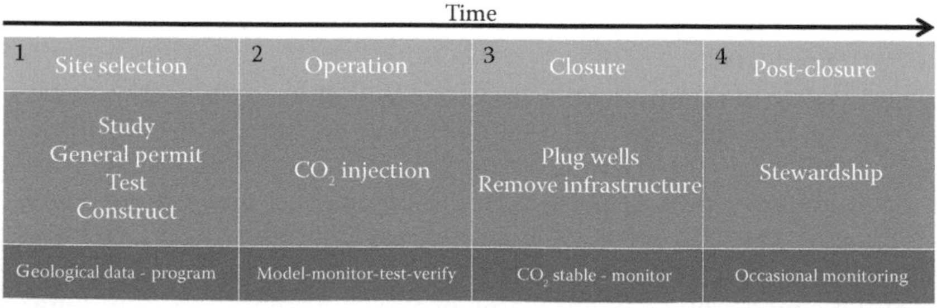

Time			
1 Site selection	**2** Operation	**3** Closure	**4** Post-closure
Study General permit Test Construct	CO$_2$ injection	Plug wells Remove infrastructure	Stewardship
Geological data - program	Model-monitor-test-verify	CO$_2$ stable - monitor	Occasional monitoring

FIGURE 1.2
Phases involved in a carbon storage project.

10% and 15%. Additionally, the minimum required formation thickness for storage sites is about 20 m. Another storage characteristic for a favorable CO$_2$ storage site is a continuous sealing system that caps the reservoir in order to prevent fluid flow in the upward direction (leakage). Prior to injection, the integrity of the seal should be verified to assure secure storage containment. "Injectivity," which represents the rock/fluid capability for CO$_2$ flow in the reservoir, is another storage site characteristic. The ideal permeability values for CO$_2$ storage sites are more than 9.87×10^{-14} m^2 (100 mD) (4).

1.2 Numerical Reservoir Simulation

Reservoir simulation models are the main tools used to perform the principal studies related to uncertainty analysis of CO$_2$ sequestration. They provide a means to predict the performance of the injection and storage process under different geological realizations and different injection scenarios. The outcome of any realization can be studied for a number of different aspects:

- Distribution of CO$_2$ though the pore spaces of the formation and CO$_2$ front prediction
- Displacement of water, oil, or gas caused by CO$_2$ injection
- Pressure build-up due to injection

In order to comprehensively study and quantify the risk associated with a sequestration project, thousands (and sometimes tens or hundreds of thousands) of simulations runs are required to take into account the variability of the uncertain parameters in the project, which can be used practically for an extensive uncertainty analysis.

In GCS, once the CO$_2$ is injected underground, there is no direct method through which the real CO$_2$ flow path in the porous media can be determined. Numerical reservoir simulation models are the major tools through which the fate of the CO$_2$ can be studied after being injected underground. These models are constructed based on geological studies and interpretations, field observation, and measurements, and therefore are essentially uncertain. In each specific CO$_2$ storage project, different operational practices will result in different storage outcomes. Consequently, any practical uncertainty analysis and risk assessment technique should address both geological and operational uncertainties.

Developing a reservoir simulation model and executing it tens of thousands of times under different operational scenarios and different distributions of reservoir characteristics ultimately provides the required solution space that can be used for a practical uncertainty analysis. The most widely used technique for this purpose is Monte Carlo simulation. However, accomplishing uncertainty analysis using the Monte Carlo simulation method is impractical for a geologically complex reservoir simulation model, which requires a few hours (or even several minutes) for a single run (execution).

On the one hand, adding complexity to the reservoir simulation model is inevitable because integrating all the observations and measurements is the sensible way to reduce uncertainty. On the other hand, the more complex the simulation model, the higher the run (execution) time. Therefore, any study which involves thousands of simulation runs, such as uncertainty analysis, an optimization study, or history matching can become prohibitively long (impractical) due to the massive requirements of run time and computational effort.

The long execution times of numerical reservoir simulation models have been a problem that the oil and gas industry has been facing for a very long time. This is actually true for any industry that relies on numerical solutions to complex physics represented by partial differential equations. Computational fluid dynamics (CFD) is another example. To address the execution (run) time problem associated with numerical simulations, scientists and engineers have devised a set of solutions that in general are referred to as "proxy modeling." The idea is to reduce the computational costs associated with complex numerical simulations. Of course, as long as the process of proxy modeling is performed within the context of the computational sciences (numerical solutions of complex partial differential equations) as a scientific paradigm, then there is limited room for maneuver.

In other words, as long as we stay within the paradigm of computational sciences, it would be impossible to simultaneously capture all the complex physics of a given process, perform the analysis at high resolution in space and time (enough to accommodate the complexity of the process), and perform all the required computation and achieve the desired results in a short period of time. Traditional proxy modeling forces scientists and engineers to choose between accuracy and speed. While accuracy compromises the speed at which the results are achieved, speed will compromise the accuracy.

In order to address this problem, efforts have been made to develop proxy models which can be used as a substitute for a complex reservoir simulation models. What are sought are proxy models that can generate a meaningful representation of the existing complex system and which can be performed in a computationally efficient way. It was soon learnt that it is impossible to accomplish this objective within the context of the computational sciences. The proxy models that were developed would sacrifice one of the two required characteristics for practical uncertainty analysis or optimization. Fast proxy models must either simplify the physics being modeled or significantly reduce the resolution in space and time.

2

Petroleum Data Analytics

Shahab D. Mohaghegh

CONTENTS

The realization that much value can be extracted from the data routinely collected (much of it is left unused) during drilling, completion, stimulation, workover, injection, and production operations in the upstream exploration and production (E&P) industry has resulted in a growing interest in the application of data-driven analytics in our industry. Related activities that had been regarded as exotic academic endeavors have now come to the forefront and are attracting much attention. New startups are entering the marketplace, some with good products and expertise, and others purely based on marketing gimmicks and opportunistic intuitions.

Petroleum data analytics (PDA) is the application of data-driven analytics and big data analytics in the upstream oil and gas industry. It is the application of a combination of techniques that make the most of the data collected in the oil and gas industry in order to analyze, model, and optimize production operations. Since the point of departure for this technology is data rather than physics and geology, it provides an alternative to the conventional solutions that have been used in the industry for the past century. Petroleum data analytics lends itself, very favorably to numerical simulation and modeling of carbon storage in geological formations since very large amounts of data can be generated from such numerical models.

The techniques that are incorporated in the author's work to perform the types of analysis, modeling, and optimization that are presented in this book include artificial intelligence, machine learning and data mining, and, more specifically, artificial neural networks, evolutionary optimization, fuzzy set theory. The following sections in this chapter provide some brief explanations on each of these techniques. Much more can be learned about these technologies by digging deeper into the references that are provided at the end of the book.

2.1 Artificial Intelligence

Artificial intelligence (AI) is the area of computer science focusing on creating machines that can engage in behaviors that humans consider intelligent. The ability to create intelligent machines has intrigued humans since ancient times, and today, with the advent of the computer and 50 years of research into AI programming techniques, the dream of smart machines is becoming a reality. Researchers are creating systems which can mimic human thoughts, understand speech, beat the best human chess player, win challenging "Jeopardy" contests, and countless other feats never before possible.

AI is a combination of computer science, physiology, and philosophy. It is a broad topic, consisting of different fields, from machine vision to expert systems. The element that the fields of AI have in common is the creation of machines that can "think." In order to classify machines as "thinking," it is necessary to define intelligence. To what degree does intelligence consist of, for example, solving complex problems, or making generalizations and relationships? And what about perception and comprehension?

AI may be defined as a collection of analytic tools that attempts to imitate life. AI techniques exhibit an ability to learn and deal with new situations. Artificial neural networks, evolutionary programming, and fuzzy logic are among the paradigms that are classified as AI. These techniques possess one or more attributes of "reason," such as generalization, discovery, association, and abstraction. In the last two decade AI has matured into a set of analytic tools that facilitate solving problems that were previously difficult or impossible to solve. The trend now is the integration of these tools, as well as with conventional tools such as statistical analysis, to build sophisticated systems that can solve challenging problems. These tools are now used in many different disciplines and have found their way into commercial products. AI is used in areas such as medical diagnosis, credit card fraud detection, bank loan approval, smart household appliances, automated subway system controls, self-driving cars, automatic transmissions, financial portfolio management, robot navigation systems, and many more.

In the oil and gas industry these tools have been used to solve problems related to drilling, reservoir simulation and modeling, pressure transient analysis, well log interpretation, reservoir characterization, and candidate well selection for stimulation, among other things.

2.2 Data Mining

As the volume of data increases, human cognition is no longer capable of deciphering important information from it by conventional techniques. Data mining and machine learning techniques must be used in order to deduce information and knowledge from the raw data that resides in the databases.

Data mining is the process of extracting hidden patterns from data. With a marked increase in the amount of data that is being routinely collected, data mining is becoming an increasingly important tool to transform collected data into information. Although the incorporation of data mining in the E&P industry is relatively new, it has been commonly used in a wide range of applications, such as marketing, fraud detection, and scientific discovery. Data mining can be applied to datasets of any size. However, while it can be used to uncover hidden patterns in the data that has been collected, obviously it can neither uncover patterns that are not already present in the data, nor uncover patterns in the data that has not been collected.

Data mining (sometimes referred to as knowledge discovery in databases, or KDD) has been defined as "the nontrivial extraction of implicit, previously unknown, and potentially useful information from data." It uses artificial intelligence, machine learning, statistical, and visualization techniques to discover and present knowledge in a form which is easily comprehensible to humans.

Data mining has been able to attract the attention of many in the fields of scientific research, business, the banking sector, intelligence agencies, and many others from the early days of its inception. However, its use has not always been as easy as it is now. Data mining is used by businesses to improve marketing and to understand the buying patterns of clients. Attrition analysis, customer segmentation, and cross selling are the most important ways through which data mining is showing new ways in which businesses can multiply their revenue.

Data mining is used in the banking sector for credit card fraud detection by identifying the patterns involved in fraudulent transactions. It is also used to reduce credit risk by classifying a potential client and predicting bad loans. Data mining is used by intelligence agencies such as the FBI and CIA to identify threats of terrorism. After the 9/11 incident, data mining has become one of the prime means with which to uncover terrorist plots.

In the popular article, "IT Doesn't Matter" (5) Nicholas Carr argued that the use of IT is nowadays so widespread that any particular organization does not have any strategic advantage over the others due to the use of IT. He concludes that IT has lost its strategic importance. However, it is the view of the author that in today's E&P landscape data mining has become the sort of tool that can provide a strategic and competitive advantage for those that have the foresight to embrace it in their day-to-day operations. It is becoming more and more evident that NOCs (National Oil Companies) and IOCs (International Oil Companies), and independent operators can create strategic advantages over theirs competitors by making use of data mining to obtain important insights from the collected data.

2.2.1 Steps Involved in Data Mining

There are various steps that are involved in mining data, including the following:

1. *Data integration*: The reality is that in today's oil and gas industry data is never in the form that one needs in order to perform data mining. Usually there are multiple sources for data, and the data exists in several databases. Data needs to be collected and integrated from different sources in order to be prepared for data mining.

2. *Data selection*: Once the data is integrated it is usually used for a specific purpose. The collected data needs to be studied and the specific parts of the data that lend themselves to the task at hand in an organization should be selected for the given data mining project. For example, human resources data may not be needed for a drilling project.

3. *Data cleansing*: The data that has been collected is usually not clean and may contain errors, missing values, noise, or inconsistencies. Different techniques need to be applied to the selected data in order to get rid of such anomalies.

4. *Data abstraction and summarization*: The data that has been collected, especially if it is operational data, may need to be summarized while keeping the main essence of its behavior intact. For example, if pressure and temperature data is collected via a permanent down-hole gauge at one-second frequency, it may need to be summarized or abstracted into minute data before use in a specific data-mining project.

5. *Data transformation*: Even after cleaning and abstraction, data may not be ready for mining. Data usually needs to be transformed into forms appropriate for mining. The techniques used to accomplish this include smoothing, aggregation, normalization, and so on.

6. *Data mining*: Machine learning and techniques such as clustering and association analysis are among the many different techniques used for data mining. Both descriptive and predictive data mining may be applied to the data. The objectives of the project will determine the type of data mining and the techniques used in it.

7. *Pattern evaluation and knowledge presentation*: This step involves visualization, transformation, and removing redundant patterns from the patterns generated.

8. *Decisions/Use of discovered knowledge*: Use of the acquired knowledge to make better decisions is the essence of this last step.

2.3 Artificial Neural Networks

Much has been written about artificial neural networks. There are books and articles that can be accessed for a deep understanding of the topic and all relevant algorithms. The objective here is to provide a brief overview, sufficient to enable an understanding of the topics presented in this book, and make them easy to follow. For a more detailed understanding of this technology, it is highly recommended that the reader refer to the books and articles referenced here.

Neural network research can be traced back to a paper by McCulloch and Pitts in 1943 (6). In 1958 Frank Rosenblatt invented the "Perceptron" (7). Rosenblatt proved that, given linearly separable classes, a perceptron will, in a finite number of training trials, develop a weight vector that will separate the classes (a pattern classification task). He also showed that his proof holds independent of the starting value of the weights. Around the same time Widrow and Hoff (8) developed a similar network called "Adeline." Minsky and Papert (9) in a book called "Perceptrons" pointed out that the theorem obviously applies to those problems that the structure is capable of computing. They showed that elementary calculations such as simple "exclusive or" (XOR) problems cannot be solved by single-layer Perceptrons.

Rosenblatt (7) also studied structures with more layers and believed that they could overcome the limitations of simple Perceptrons. However, there was no learning algorithm known that could determine the weights necessary to implement a given calculation. Minskey and Papert doubted that one could be found and recommended that other approaches to artificial intelligence should be pursued. Following this discussion, most of the computer science community left the neural network paradigm for twenty years (10). In the early 1980s Hopfield was able to revive neural network research. Hopfield's efforts coincided with development of new learning algorithms such as back-propagation. The growth of neural network research and applications has been phenomenal since this revival.

2.3.1 Structure of a Neural Network

An artificial neural network is an information-processing system that has certain performance characteristics in common with biological neural networks. Therefore, it is appropriate to briefly describe a biological neural network before offering a detailed definition of artificial neural networks.

All living organisms are made up of cells. The basic building blocks of the nervous system are nerve cells, called neurons. Figure 2.1 shows a schematic diagram of two bipolar neurons. A typical neuron contains a cell body where the nucleus is located, dendrites, and an axon. Information in the form of a train of electro-chemical pulses (signals) enters the cell body from the dendrites. Based on the nature of this input the neuron will activate in an excitatory or inhibitory fashion, and provides an output that will travel through the axon and connects to other neurons where it becomes the input to the receiving neuron. The point between two neurons in a neural pathway, where the termination of the axon of one neuron comes into close proximity with the cell body or dendrites of another, is called a synapse. Signals traveling from the first neuron initiate a train of electro-chemical pulses (signals) in the second neuron.

It is estimated that the human brain contains on the order of 10–500 billion neurons (11). These neurons are divided into modules, and each module contains about 500 neural networks (12). Each network may contain about 100,000 neurons, where each neuron is connected to hundreds to thousands of other neurons. This architecture is the main driving force behind the complex behavior that comes so natural to us. Simple tasks such as catching a ball, drinking a glass of water, or walking in a crowded market require so many complex and coordinated calculations that sophisticated computers are unable to undertake the task, and yet it is done routinely by humans without a moment of thought.

This becomes even more interesting when one realizes that neurons in the human brain have a cycle time of about 10–100 milliseconds, while the cycle time of a typical desktop computer chip is measured in nanoseconds (about 10 million times faster than the human brain). The human brain, although a million times slower than common desktop PCs, can perform many tasks orders of magnitude faster than computers because of its massively parallel architecture.

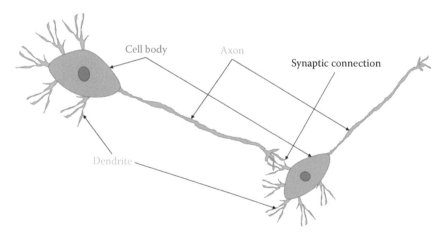

FIGURE 2.1
Schematic diagram of two bipolar neurons.

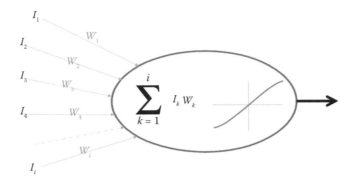

FIGURE 2.2
Schematic diagram of an artificial neuron or a processing element.

Artificial neural networks are a rough approximation and simplified simulation of the process explained above. An artificial neural network can be defined as an information-processing system that has certain performance characteristics similar to biological neural networks. They have been developed as a generalization of mathematical models of human cognition or neural biology, based on the following assumptions:

1. Information processing occurs in simple processing elements, called neurons.
2. Signals are passed between neurons over connection links.
3. Each connection link has an associated weight, which, in a typical neural network, multiplies the signal being transmitted.
4. Each neuron applies an activation function (usually non-linear) to its net input to determine its output signal (13).

Figure 2.2 is a schematic diagram of a typical neuron (processing element) in an artificial neural network. The output from other neurons is multiplied by the weight of the connection and enters the neuron as input. Therefore an artificial neuron has many inputs and only one output. The inputs are summed and subsequently applied to the activation function, and the result is the output of the neuron.

2.3.2 Mechanics of Neural Networks Operation

An artificial neural network is a collection of neurons that are arranged in specific formations. Neurons are grouped into layers. In a multilayer network there is usually an input layer, one or more hidden layers and an output layer. The number of neurons in the input layer corresponds to the number of parameters that are being presented to the network as input. The same is true for the output layer. It should be noted that neural network analysis is not limited to a single output and that neural networks can be trained to build data-driven models with multiple outputs. The neurons in the hidden layer or layers are mainly responsible for feature extraction.

An increase in the number of hidden neurons provide increased dimensionality and accommodate tasks such as classification and pattern recognition. Figure 2.3 is a schematic diagram of a fully connected three-layer neural network. There are many kinds of neural network. Neural network scientists and practitioners have provided different classifications to describe these. One of the most popular classifications is based on training methods, whereby neural networks can be divided into two major categories, namely supervised and unsupervised neural networks. Unsupervised neural networks, also known as self-organizing maps, are mainly clustering and classification algorithms. They have been used in the oil and gas industry to interpret well logs and to identify lithology. They are called unsupervised simply because no feedback is provided to the network by the person that is training the neural network. The network is asked to classify the input vectors into groups and clusters. This requires a certain degree of redundancy in the input data and hence the notion that redundancy is knowledge (14).

Most of the practical and useful neural network applications in the upstream oil and gas industry are based on supervised training algorithms. During a supervised training process both input and output are presented to the network to permit learning on a feedback basis. A specific architecture, topology, and training algorithm are selected and the network is trained until it converges to an acceptable solution. During the training process, the neural network tries to converge to an internal representation of the system behavior. Although by definition neural networks are model-free function approximators, some people choose to call the trained network a "neuro-model." In this book our preferred terminology is "data-driven" model.

The connections correspond roughly to the axons and synapses in a biological system, and they provide a signal transmission pathway between the nodes. Several layers can be interconnected. The layer that receives the inputs is called the input layer. It typically performs no function other than buffering

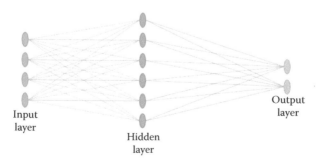

FIGURE 2.3
Schematic diagram of a three-layer neuron network.

of the input signal. In most cases, the calculation performed in this layer is a normalization of the input parameters so that parameters such as porosity (which is usually represented as a fraction) and initial pressure (usually in thousands of psi) would be treated equally by the neural network at the start of the training process. The network outputs are generated from the output layer. Any other layers are called hidden layers because they are internal to the network and have no direct contact with the external environment. Sometimes they are likened to a "black box" within the network system. However, just because they are not immediately visible does not mean that one cannot examine the function of those layers. There may be zero to several hidden layers in a neural network. In a fully connected network every output from one layer is passed along to every node in the next layer.

In a typical neural data-processing procedure, the database is divided into three separate portions called training, calibration, and verification sets. The training set is used to develop the desired network. In this process (depending on the training algorithm that is being used), the desired output in the training set is used to help the network adjust the weights between its neurons or processing elements. During the training process the question arises of when to stop the training? How many times should the network go through the data in the training set in order to learn the system behavior? When should the training stop? These are legitimate questions, because a network can be over trained. In the neural-network-related literature, over-training is also referred to as memorization. Once the network memorizes a dataset, it would be incapable of generalization. It will fit the training dataset quite accurately, but suffers in generalization. The performance of an over-trained neural network is similar to a complex non-linear regression analysis.

Over-training does not apply to some neural network algorithms simply because they are not trained using an iterative process. Memorization and over-training are applicable to those networks that are historically among the most popular ones for engineering problem solving. These include back-propagation networks that use an iterative process during training.

In order to avoid over-training or memorization, it is a common practice to stop the training process every so often and apply the network to the calibration dataset. Since the output of the calibration dataset is not presented to the network (during the training), one can evaluate the network's generalization capabilities by how well it predicts the calibration set's output. Once the training process is completed successfully, the network is applied to the verification dataset.

During the training process each artificial neuron (processing element) handles several basic functions. First, it evaluates input signals and determines the strength of each one. Second, it calculates a total for the combined input signals and compares that total to some threshold level. Finally, it determines what the output should be. The transformation of the input to output – within a neuron – takes place using an activation function. Figure 2.4 shows two of the commonly used activation (transfer) functions.

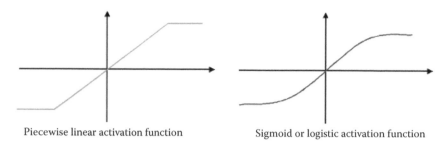

Piecewise linear activation function   Sigmoid or logistic activation function

FIGURE 2.4
Commonly used activation functions in artificial neurons.

All the inputs come into a processing element (in the hidden layer) simultaneously. In response, the neuron either "fires" or "doesn't fire," depending on some threshold level. The neuron will be allowed a single output signal, just as in a biological neuron – many inputs, one output. In addition, just as things other than inputs affect real neurons, some networks provide a mechanism for other influences. Sometimes this extra input is called a bias term, or a forcing term. It could also be a forgetting term, when a system needs to unlearn something (15).

Each input is initially assigned a random relative weight (in some advanced applications – based on the experience of the practitioner – the relative weight assigned initially may not be random). During the training process the weight of the inputs is adjusted. The weight of the input represents the strength of its connection to the neuron in the next layer. The weight of the connection will affect the impact and influence of that input. This is similar to the varying synaptic strengths of biological neurons. Some inputs are more important than others in the way they combine to produce an impact. Weights are adaptive coefficients within the network that determine the intensity of the input signal. The initial weight for a processing element could be modified in response to various inputs and according to the network's own rules for modification.

Mathematically, we could look at the inputs and the weights on the inputs as vectors, such as $I_1, I_2, I_3, I_4 \ldots I_n$ for inputs and as $W_1, W_2, W_3, W_4 \ldots W_n$ for weights. The total input signal is the dot, or inner, product of the two vectors. Geometrically, the inner product of two vectors can be considered a measure of their similarity. The inner product is at its maximum if the vectors point in the same direction. If the vectors point in opposite directions (180°), their inner product is at its minimum.

Signals coming into a neuron can be positive (excitatory) or negative (inhibitory). A positive input promotes the firing of the processing element, whereas a negative input tends to keep the processing element from firing. During the training process some local memory can be attached to the processing element to store the results (weights) of previous computations. Training is accomplished by modification of the weights on a continuous basis until convergence is reached. The ability to change the weights allows the network to modify its behavior in response to its inputs, or to learn. For example, suppose a network is being trained to correctly calculate the initial production from a newly drilled well. At the early stages of the training the neural network calculates the initial production from the new well to be 150 bbls/day, whereas the actual initial production is 1000 bbls/day. On successive iterations (training), connection weights that respond to an increase in initial production (the output of the neural network) are strengthened, and those that respond to a decrease are weakened until they fall below the threshold level and the correct calculation of the initial production is achieved.

In the back-propagation algorithm (16) (one of the most commonly used supervised training algorithms in upstream oil and gas operations) the network output is compared with the desired output – which is part of the training dataset – and the difference (error) is propagated backward through the network. During this back-propagation of error the weights of the connections between neurons are adjusted. This process is continued in an iterative manner. The network converges when its output is within acceptable proximity of the desired output.

2.3.3 Practical Considerations during the Training of a Neural Network

There is a substantial amount of art involved in the training of a neural network. In this section of the book the objective is to share some personal experiences that have been gained throughout many years of developing data-driven models for oil- and gas-related problems. These are mainly practical considerations and may or may not agree with similar practices in other industries, but they have worked very well for the author over the past two decades.

Understanding how machines learn is not complicated. The mathematics associated with neural networks, including how they are built and trained, is not really complex (16). It includes vector calculus and the differentiation of some normal functions. However, being a good reservoir or production engineer and a capable petroleum engineering modeler is essential for building and effectively utilizing the data-driven models that are covered in this book. In order to be a capable petroleum data scientist, while it is necessary to be a competent petroleum engineer, it is not a requirement to have degrees in mathematics, statistics, or machine learning.

In order to develop functional data-driven models there is no need to be an expert in machine learning or artificial neural networks. What is required, though, is an ability to understand the fundamentals of this technology and eventually become an effective user and practitioner of it. Although such skills are not taught as part of any petroleum engineering curriculum at universities, acquiring such skills is not a far-reaching task, and any petroleum engineer with a bachelor's degree should be able to master it with some training and effort. It is important to understand and subscribe to the philosophy of machine learning. This means that, although as an engineer you have learned to solve problems in a particular way, you need to understand and accept that there is more than one way to solve engineering-related problems.

The technique that engineers have used to solve problems follows a well-defined path of identifying the parameters involved and then constructing the relationships between the parameters (using mathematics) to build a model. The philosophy of solving a problem using machine learning is completely different. Given that in machine-learning algorithms such as artificial neural networks, models are built using data, the path to follow in order to solve an engineering problem differs from that you learned as a petroleum engineer. To solve problems using data, you have to be able to teach an open computer algorithm about the problem and its solutions. This process is called supervised learning. You have to create (actually collect) a large number of records (examples) that include the inputs (parameters involved) and the outputs (the solution you are trying to solve for). During training, these records (the coupled input–output pairs) are presented to the machine-learning algorithm, and by repetition (redundancy sometimes plays a positive role in this process) the machine will eventually learn how the problem is solved. The algorithm does this by building an internal representation of the mapping between inputs and outputs.

In the previous section of this book, the fundamentals of this technology were covered. In this section, some practical aspects of this technology will be briefly discussed. These practical aspects will help a data-driven modeler learn how to train good and useful neural networks. The neural network training includes several steps, which are also covered here.

2.3.3.1 Selection of Input Parameters

Since all models are wrong, the scientist cannot obtain a "correct" one by excessive elaboration. On the contrary, following William of Occam,* the scientist should seek an economical description of natural phenomena. Just as the ability to devise simple but evocative models is the signature of the great scientist, over-elaboration and over-parameterization are often the mark of mediocrity (17). In order not to over-parameterize our neural network model we need to use the right number of variables.

Selection of the input parameters that are going to be used to train a neural network from the variables (potential inputs) that have been assimilated in the database is not a trivial procedure. A database generated for a given project usually includes a very large number of parameters, all of which are potential input parameters to the neural networks that are going to be trained for the data-driven model. They include static parameters and dynamic parameters,

---

* Occam's razor is a problem-solving principle devised by William of Ockham (c. 1287–1347). The principle states that, among competing hypotheses, the one with the fewest assumptions should be selected. In the absence of certainty, the fewer assumptions that are made, the better.

as well as similar parameters for several offset wells. These parameters are designated as columns in a flat file that is eventually used to train data-driven models (neural networks).

Not all of the parameters that are included in the database are used to train the neural networks. Actually, it is highly recommended that the number of parameters that are used to build (train, calibrate, and validate) the neural network be limited. This limitation of the input parameters should not be interpreted as the other parameters not playing any role in forming or calculating the output of the model. The elimination of some of the parameters and the use of some others to build the data-driven model simply means that a subset of the parameters play such an important role in the determination of the model output that they overshadow (or sometimes implicitly represent) the impact of other parameters. In other words a model can be developed using only these parameters and safely ignoring others.

Only a subset of these parameters should therefore be selected and used as the input to the data-driven models. Experience with developing successful data-driven models has shown that the process of selecting the parameters that must be used as input to the model needs to satisfy the following three criteria:

- The impact (influence) of all the exiting parameters (in the database) on the model output should be identified and ranked. Then the top "x" percent of these ranked parameters should be used as input in the model. This is easier said than done. There are many techniques that can be used to help the data-driven modeler in identifying the influence of parameters on a selected output. These techniques can be as simple as linear regression and as complex as fuzzy pattern recognition.* Some use principal component analysis (18) to accomplish this task.

- In the list of input parameters that are identified to be used in the training of the neural network, there must exist parameters that can validate the physics and/or the geology of the model. If such parameters are already among the highly ranked parameters in the previous step, then great, otherwise, the data-driven modeler must see to it that they are included in the model. Being able to verify that the data-driven model has understood the physics and honors it is an important part of data-driven modeling.

- In many cases the data-driven model is developed in order to optimize production. Identification of optimized choke setting during production is a good example of such a situation. In such cases, parameters that are needed to optimize production should be included in the set of input parameters. If the optimization

---

* This is a proprietary algorithm developed by Intelligent Solutions, Inc. and used in their Data Driven Modeling software application called IMprove™ (www.IntelligentSolutionsInc.com).

parameters are already among the highly ranked parameters in the previous step, then great, otherwise, the data-driven modeler must see to it that they are included in the model.

The machine learning literature includes many techniques for this purpose. In these publications the technology is referred to as "feature selection."

2.3.3.2 Partitioning the Dataset

Data in the spatio-temporal database is transferred into a flat file once a superset of parameters are selected to be used in the training of the neural networks. The data in the flat file needs to be partitioned into three segments: training, calibration, and validation. As will be discussed in the next section, the way these segments are treated determines the essence of the training process. In this section, the characteristics of each of these data segments and their use and purpose are briefly discussed.

In general, the largest of the three segments is the training dataset. This is the data that is used to train the neural network and create the relationships between the input parameters and the output parameter. Everything that one wishes to teach a data-driven model must be included in the training dataset. One must realize that the range of the parameters as they appear in the training set determines the range of the applicability of the data-driven model. For example, if the range of permeability in the training set is between 2 and 200 mD, one should not expect the data-driven model to perform reasonably well for a data record with permeability values less than 2 mD and higher than 200 mD. This is due to the well-known fact that most machine-learning algorithms, neural networks included, demonstrate great interpolative capabilities, even if the relationship between the input parameters and the output(s) is highly non-linear. However, machine-learning algorithms are not known for their extrapolative capabilities.

As we mentioned in Section 2.3.1, the input parameters are connected to the output parameter through a set of hidden neurons. The strength of the connections between neurons (between input neurons and hidden neurons, between hidden neurons with one another if such connections exists, and between hidden neurons and output neurons) is determined by the weight associated with each connection. During the training process, the optimum weight of each connection is determined through an iterative learning process.

During the training process, the weights between the neurons (also known as synaptic connections) in a neural network find their optimum value. Collection of these optimum values forms the coefficient matrices that are used to calculate the output parameter. Therefore, the role of the training dataset is to help the modeler determine the strength between the neurons in a neural network. Convergence of a network to a desirable set of weights that will translate to a well-trained and smart neural network model depends on the information content of the training dataset. When the neural networks are being trained for data-driven model purposes, the size of the training dataset may be as high as 80% or as low as 40% of the entire dataset. This percentage is a function of the number of records in the database.

The calibration dataset is not used directly during the training process, and it actually plays no direct role in changing the weights of the connections between the neurons. The calibration dataset is a blind dataset that is used after every epoch of training* in order to test the quality and the goodness of the trained neural network. In many circles this is also called the "test" set. The calibration dataset is essentially a watch dog that observes the training process and decides when to stop the training process, because the network is only as good as its prediction of the calibration dataset (a randomly selected dataset that is actually a blind dataset).

Therefore, after every epoch of training (when the network gets to see all the records in the training dataset, once) the weights are saved and the network is tested against the calibration dataset to see if there has been any improvement of network predictive performance against this blind dataset. This test of network predictive capabilities is performed after every epoch to monitor its generalization capabilities. Usually one or more metrics such as R^2, the correlation coefficient, or the mean square error (MSE) are used to calculate the network's generalization capabilities. These metrics are used to determine how closely the set of synaptic connection weights will enable the calculation of the outputs as a function of the input parameters by comparing the output values computed by the neural networks against those measured in the field (the actual or real outputs) and used to train the neural network.

As long as this metric is improving for the calibration dataset, it means that the training can continue and the network is still learning. When the neural networks are being trained for data-driven modeling purposes, the size of the calibration dataset is usually between 10% and 30% of the entire dataset, depending on the size of the database.

The last, but arguably the most important, dataset (segment) is the validation or verification dataset. This dataset plays no role during training or calibration of the neural network. It has been selected and put aside from the very beginning to be used as a blind dataset. It literally sits on the sidelines and does nothing until the training process is over. This blind dataset validates the generalization capabilities of the trained neural network.

While having no role to play during the training and calibration of the neural network, this dataset validates the robustness of the predictive capabilities of the neural network. The data-driven model that will result from the neural network that is being trained is as good as the outcome of the validation or verification dataset. When neural networks are being trained for data-driven modeling purposes, the size of the validation (verification) dataset is usually between 10% and 30% of the entire dataset, depending on the size of the database.

---

* An epoch of training is completed when all the data records in the training set have been passed through the neural network and the error between neural network output and the actual field measurements are calculated.

Since the database is being partitioned into three datasets, it is important to make sure that the information content of each dataset is comparable to the others. If they differ, and often they will, then it would be best that the training set have the largest, most comprehensive information content of the three datasets. This will ensure a healthy training behavior and will increase the likelihood of training a good and robust neural network. Information content and its relationship with entropy within the context of information theory is an interesting subject that those involved with data-driven analytics and machine learning should understand (19).

2.3.3.3 Structure and Topology

The structure and topology of a neural network is determined by several factors, and hypothetically can have an infinite number of possible forms. However, almost all of them include a combination of factors, such as the number of hidden layers, the number of hidden neurons in each hidden layer, the combination of the activation functions, and the nature of the connections between neurons. In this section the objective is to briefly discuss some of the most popular structures and specifically those that have shown success when used in the development of data-driven models for oil- and gas-related applications. In other words, the intention is not to turn this chapter of the book into a neural network tutorial, but rather to present practices that have proven successful during the development of data-driven models in the past for the author and can be used as a "rule of thumb" for those entering the world of data-driven modeling.

As far as the connection between neurons is concerned, the structures that have been used most successfully in data-driven models are fully connected neural networks. In fully connected networks every input neuron is connected to every hidden neuron, and every hidden neuron is also connected to the output neuron; this network is called a fully connected network, as shown in Figures 2.5 through 2.7.

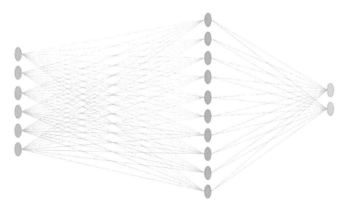

FIGURE 2.5
A fully connected neural network with one hidden layer that includes 11 hidden neurons, seven input neurons and one output neuron.

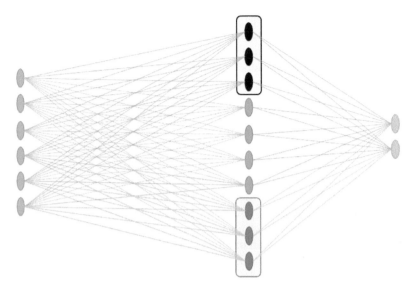

FIGURE 2.6
A fully connected neural network with one hidden layer that includes three different sets of activation functions along its 11 hidden neurons, seven input neurons and one output neuron.

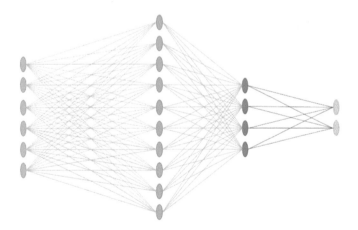

FIGURE 2.7
A fully connected neural network with two hidden layer that includes 11 and seven hidden neurons, respectively, as well as seven input neurons and one output neuron.

Figure 2.5 shows the most simple and also the most popular type of neural network for the development of data-driven models that form the main engines of the data-driven model. This is a simple, three-layer, fully connected neural network. The three layers are the input layer, the hidden layer, and the output layer. Furthermore, while the number of output neurons in the output layer can be more than one, our experience with data-driven models has shown that, except in some specific situations, a single output neuron in the output layer performs best.

Furthermore, after experience with a large number and variety of network structures, the author's experience has shown that if you are not able to train a good network* using a simple structure as shown in Figure 2.5, your chances of achieving a good neural network with any other structure will be slim. In other words, it is not the structure of the neural network that will make or break the success of your efforts in training a data-driven model, rather it is the quality and information content of the database that determines your chances of being successful in developing a neural network, and by the same token, a data-driven model. If you see persistent issues in developing the data-driven model (the networks cannot be trained properly or do not have good predictive capabilities), you need to revisit your database rather than playing around with the structure and the topology of your neural network.

In the author's opinion, when practitioners of machine learning (specifically in the upstream oil and gas industry) concentrate their efforts, and naturally the presentation of their data-driven modeling efforts, on the neural network structure and how it can be modified in order to control the training of the model, one must take that as an indication of naivety and a lack of substance and skill of the practitioner in machine learning technology, and not an indication of expertise. You will find that an expert in the art and science of machine learning will concentrate most on the information content and details of the dataset being used to train the neural network, as long as the very basic issues (the number of hidden layers and hidden neurons, the learning rate and momentum, etc.) of the structure of the neural network have been reasonably determined to be solid. The details of the structure and topology of a neural network can have an enhancing impact on the results, but it will not make or break your data-driven model.

Sometimes, changing some of the activation functions can help in fine-tuning a neural network's performance, as shown in Figure 2.6. In this figure the hidden neurons are divided into three segments and each set of hidden neurons can be assigned a different activation function. Some details about activation functions were covered in previous sections. The author does not recommend that the initial structure of a neural network model be designed as shown in Figure 2.6; rather, if need be, the structure in Figures 2.6 and 2.7 may be used to enhance the performance of a network that has shown serious promise to be a good network, but its enhancement requires some fine-tuning.

Once the structure of the neural network has been determined, it is time to decide upon the learning algorithm. By far the most popular learning (training) algorithm is "error back-propagation" (16) (or simply referred to as "back-propagation"). In this learning algorithm the network calculates

---

* What constitutes a "good network"? A good network is a network that can be trained and calibrated and validated. It can learn well and has robust predictive capabilities. The rest is very problem dependent.

a series of outputs based on the current values of the weights (strength of the connections between neurons – synaptic connections) and compares its calculated outputs for all the records with actual (measured) outputs (what it is trying to match).

The calculated error between the network output and the measured values (also known as the target) is then back-propagated throughout the network structure with the aim of modifying the synaptic weights between neurons as a function of the magnitude of the calculated error. This process is continued until the back-propagation of the error and modification of the connection weights no longer enhances the network performance.

Several parameters are involved and can be modified during this training process to impact the progression of the network training. These parameters include the network's learning rate and the momentum for the weights between each set of neurons (layers), as well as the nature of the activation function. However, just as mentioned before, none of these factors will make or break a neural network; rather they can be instrumental in fine-tuning the result of a neural network. The information content of the database (which is essentially domain expertise related to petroleum engineering and geosciences) is the most important factor in the success or failure of a data-driven model.

2.3.3.4 Training Process

Since neural networks are known to be universal function approximators, hypothetically speaking they are capable of complete replication (reproduction) of the training dataset. In other words, given enough time and a large enough number of hidden neurons, a neural network should be able to reproduce the outputs of the training set from all the inputs, with 100% accuracy. This is something that one expects from a statistical approach or mathematical spline curve-fitting process. Such a result from a neural network is highly *undesirable,* and must be avoided. This is due to the fact that a neural network that is so accurate on the training set has literally memorized all the training records and has next to no predictive value.

This is the process that is usually referred to as over-training or over-fitting, and in the artificial intelligence lingo is referred to as "memorization" and must be avoided. An over-trained neural network memorizes the data in the training set and can reproduce the output values, almost identically, and does not learn it. Therefore, it cannot generalize and will not be able to predict the outcome of new data records. Such a model (if it can actually be called a model) is merely a statistical curve fit and has no value whatsoever.

Some of the geo-statistical techniques that are currently used by many commercial software applications to populate geo-cellular models are examples of such a technique (using neural networks as a curve fitting technique). Some of the most popular geo-modeling software applications that are quite cognizant of this fact have incorporated neural networks as part of their tools. However, a closer look at the way neural networks have been implemented in these software applications reveals that they are merely a marketing gimmick, and these software applications incorporate them as statistical curve-fitting techniques, which make neural network as useless as other geo-statistical techniques.

One of the roles of the calibration dataset is to prevent over-training. It is good practice to observe the network behavior during the training process in order to understand whether the neural network is in the process of converging to a solution or it needs attention from the modeler. There is much that can be learned from this observation. A simple plot of MSE versus the number of training epochs displays the neural network's training and convergence behavior. Furthermore, if the MSE is plotted for both the training dataset and the calibration dataset (after every training epoch), much can be learned from their side-by-side behavior. Several examples of such plots are shown in Figures 2.8 and 2.9.

Figures 2.8 and 2.9 include three sets of examples, each. Each example includes two graphs. The ones on the left show MSE vs. number of epochs for the calibration dataset, and the ones on the right show MSE vs. number of epochs for the training dataset. Please remember that training and calibration datasets are completely independent and have different sizes. Normally, the training dataset includes 80% and the calibration dataset includes 10% of the complete dataset. Each pair of graphs in Figures 2.8 and 2.9 represents the training process of one neural network.

During a healthy training process, the error in the calibration dataset (two-dimensional graphs on the left side of Figures 2.8 and 2.9) that is playing no role in changing the weights in the neural network is expected to behave quite similarly to the error in the training dataset (two-dimensional graphs on the right side of Figures 2.8 and 2.9). Several examples of a healthy training process are shown in Figure 2.8. In the plots shown in Figure 2.8 MSE is plotted vs. number of training epochs.

If the graphs in Figure 2.8 are plots that are being updated in real time, then the modeler can observe the error behavior of the training progress in real time and decide whether the training process should continue or should be stopped so that some modifications can to be made to the network structure or datasets. Actions such as this may be necessary once it is decided that the training process has entered a potential dead-end (lack of convergence) and there will be no more learning.

A healthy training process is defined as one where continuous, effective learning is taking place and the network is getting better with each epoch of training. One indication of such a healthy training process is the similarity of the behavior in the two plots, as shown in Figure 2.8. In the three examples shown in this figure, the error in both calibration and training sets has similar slope and behavior. This is important, because the calibration dataset is blind and independent of the training set, and such similarities in the error behavior indicate an effective partitioning of the datasets.

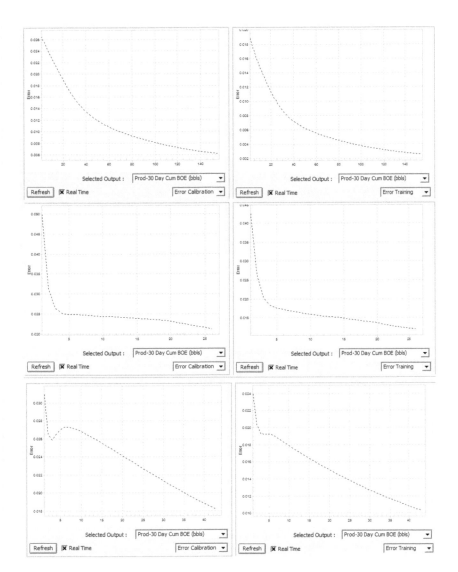

FIGURE 2.8
Plot of mean square error as a function of number of training epochs. The plot on the left shows the error for the calibration dataset and the plot on the right shows the error for the training dataset. Shown are three examples of a training process that is progressing in a satisfactory manner, when the behavior of the errors mirrors one another.

On the other hand, an unhealthy training process is one in which the behavior of the error in the training and calibration datasets differs and sometimes starts moving in opposite directions. Figure 2.9 shows three examples of unhealthy error behavior. That said, it also should be mentioned that given the nature of how gradient decent algorithms such as

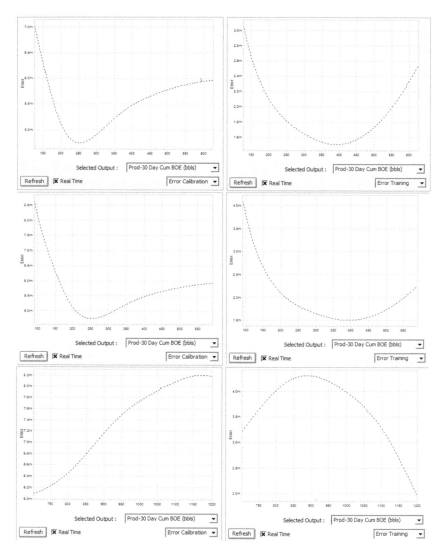

FIGURE 2.9
Plot of mean square error as a function of number of training epochs. The plot on the left shows
the error for the calibration dataset and the plot on the right shows the error for the training
dataset. Shown are three examples of a training process that are *not* progressing in a satisfactory
manner, when the behavior of the errors is different, with opposite slopes.

back-propagation work, it is expected, from time to time, that a difference
in error behavior between these two datasets will be observed, but will be
temporary. In such cases, if you give the algorithm enough time, it will correct
itself and the error behavior will begin to get healthier. This is of course a
function of the problem being solved and the prepared dataset that is being
used and must be carefully observed and judged by the modeler.

At this point in time, a legitimate question that can be asked is "What would make a training behavior unhealthy and how can it be overcome?" For example, if the best network that is saved is the one with the best (highest) R^2 and/or lowest value of MSE for the calibration dataset, how can we try to avoid an early (premature) convergence? A premature convergence is defined as a situation where the error in the training dataset is decreasing while the opposite trend is observed in the error of the calibration dataset, as shown in Figure 2.9. The answer to this question relies on the information content of the training, the calibration and the validation datasets, as mentioned in the previous section. In other words, one of the reasons why such phenomena can take place is the way the database has been partitioned. To clarify this point an example is provided. The example is demonstrated using Figures 2.10 through 2.12.

In these figures you can see that the largest value of the field measurement (*y* axis) for the output (30 days cumulative production in barrels of oil equivalent (BOE)) in the training dataset is 3850 bbls (Figure 2.10) while the largest value of the field measurement for the output in the validation dataset is 4550 bbls (Figure 2.12). Clearly, the model is not being trained on any field measurements with values larger than 3850 bbls. Therefore, the model not learning about the combined set of conditions is the reason for such a large

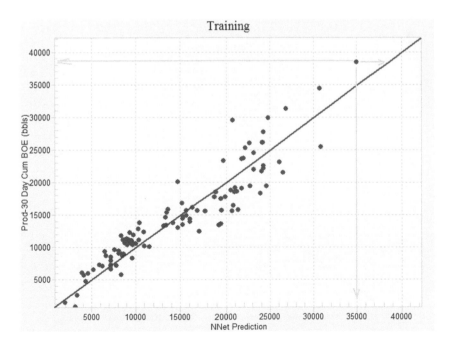

FIGURE 2.10
Cross plot of observed (measured) output versus neural network prediction for the training dataset. The largest field measurement is 3850 bbls.

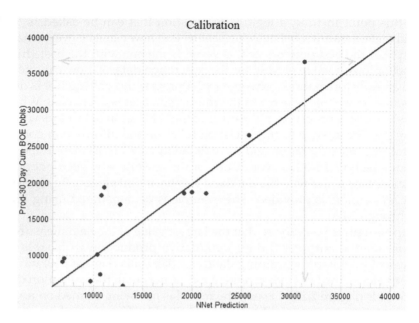

FIGURE 2.11
Cross plot of observed (measured) output versus neural network prediction for the calibration dataset. The largest field measurement is 3700 bbls.

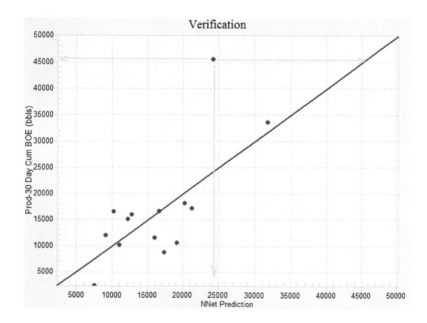

FIGURE 2.12
Cross plot of observed (measured) output versus neural network prediction for the validation dataset. The largest field measurement is 4550 bbls.

value of 30 days cumulative production. This is a clear consequence of inconsistency in the information content of the training, the calibration, and the validation datasets, which needs to be avoided.*

2.3.3.5 Convergence

In the context of data-driven models, convergence is referred to as the point when the modeler or software intelligent agent that oversees the training process decides that a better network cannot be trained and therefore the training process must end. As you may note, this is a bit different from the way convergence is defined in mathematically iterative procedures. Here, it is not advisable to identify a small enough delta error for the convergence since such an error value may never be achieved. The acceptable error in data-driven models is very much a problem-dependent issue.

In data-driven models the best type of convergence criterion is the highest R^2 or lowest MSE for the calibration dataset. It is important to note that, in many cases, these values can be a bit misleading (although they remain the best measure), and it is recommended to visually inspect the results for all the wells in the field, individually, before making a decision of whether to stop the training or continue the search for a better data-driven model.

---

* The author is not aware of any software applications for the training of neural networks that provide a means for addressing such issues. The software application that has been mentioned in one of the previous footnotes (IMprove™ by Intelligent Solutions, Inc.) is the only software application in the oil and gas industry that include a means to detect and rectify such issues. This is due to identification and then addressing of the practical issues that can be encountered when data-driven models are built to address upstream exploration and production problems.

3

Smart Proxy Modeling

Shahab D. Mohaghegh

CONTENTS

There are well-known facts regarding the shortcomings of the numerical reservoir simulation and its applications that need to be addressed if this tool is to be used effectively for the geological sequestration of CO_2. These shortcomings, which could seriously impact any studies concerning the geological storage of CO_2, can be summarized as follows:

- Numerical reservoir simulation models are uncertain by nature and their effective use for the geological storage of CO_2 requires the incorporation of technologies that allow such uncertainties to be quantified.

- Numerical reservoir simulation models are constructed for non-academic, real case scenarios that are very large (millions or tens of millions of cells). Therefore, these models have very large computational footprints. This fact limits their utilization for uncertainty analysis and quantification and for detailed field studies and optimization.

The hydrocarbon exploration and production industry has been well aware of these shortcomings, and for the past several decades a massive amount of effort has been expended to address them. The response of the industry has been to develop proxy models to address the above-mentioned issues with numerical reservoir simulation models. The proxy models that have been developed historically in the hydrocarbon exploration and production industry can be divided into the following major categories:

- *Reduced order models (ROMs)*: Numerical reservoir simulation models, especially those with formulation to model the injection, migration, and interaction of CO_2 with formation rock and native fluids in the formation, have a large amount of complex physics modeled into the simulation models. The inclusion of such complex physics sometimes requires smaller grid blocks (cells) in order to accommodate proper convergence of the numerical techniques used to solve the complex, non-linear, and higher order, partial differential equations that are involved in the fabric of such models.

  In order to accomplish their objective of reducing the computational footprint of the numerical reservoir simulation models, ROMs usually incorporate one or both of the following techniques:

  - They tend to modify/reduce the physics used to build the model;
  - They attempt to reduce the resolution in space and time. Sometimes to accomplish this they must incorporate simplification of the physics of the problem as well.

- *Statistical response surfaces (SRSs)*: Here we emphasize the word "statistical" to draw attention to the fact that the response surfaces that are developed use traditional statistics (and not machine learning) as their basis. The main difference between traditional statistics and artificial intelligence and machine learning is much deeper than some may believe and/or admit to. Although this shows itself in the techniques that are used in order to accomplish objectives, the main differences stem from the philosophical approach to problem solving. The differences are as deep seated as the differences between Aristotelian and Platonic views of the world.

3.1 Engineering Application of Data Science

Since its introduction as a discipline in the mid-90s, "data science" has been used as a synonym for applied statistics. Today, data science is used in multiple disciplines and is enjoying immense popularity. What has been causing confusion is the essence of data science as it is applied to physics-based versus non-physics-based disciplines. Such distinctions surface once data science is applied to industrial applications and when it starts moving above and beyond simple academic problems.

So what is the difference between data science as it is applied to physics-based versus non-physics-based disciplines? When data science is applied to non-physics-based problems, it is merely applied statistics. Application of data science in social networks and social media, consumer relations, demographics, or politics (some may even include medical and/or pharmaceutical sciences in this list) takes a purely statistical form, since there are no sets of governing partial differential (or other mathematical) equations that have been developed to model human behavior or the response of human biology to drugs. In such cases (non-physics-based areas), the relationship between correlation and causation cannot be resolved using physical experiments and usually, as long as they are not absurd, are justified or explained by scientists and statisticians using psychological, sociological, or biological reasoning.

On the other hand, when data science is applied to physics-based problems such as self-driving cars, or multi-phase fluid flow in reactors (computational fluid dynamics, CFD), or in porous media (reservoir simulation), it is a completely different story. The interactions between parameters that are of interest to physics-based problem solving, despite their complex nature, have been understood and modeled by scientists and engineers for decades. Therefore, treating the data that is generated from such phenomena (regardless of whether it is measurements by sensors or generated by simulation) as just numbers that need to be processed in order to learn their interactions is a gross mistreatment and over-simplification of the problem, and hardly ever generates useful results. This is why many such attempts have, at best, resulted in unattractive and mediocre outcomes. So much so that many engineers (and scientists) have concluded that data science has little serious applications in industrial and engineering disciplines.

The question may arise that if the interactions between parameters that are of interest to engineers and scientists have been understood and modeled for decades, then how could data science contribute to industrial and engineering problems? The answer is by providing a "considerable (and sometimes game changing and transformational) increase in the efficiency of the problem solving." So much so that it may change a solution from an academic exercise into a real-life solution. For example, many of the governing equations that can be solved to build and control a driverless car are well known. However, solving this complex set of non-linear equations and incorporating them into a real-time process that actually controls and drives a car in the street is beyond the capabilities of any computer today (or in the foreseeable future). Data-driven analytics and machine learning contribute significantly to accomplishing such tasks.

There is a flourishing future for data science as the new generation of engineers and scientists are exposed to and start using it in their everyday lives. The solution to clarifying and distinguishing the applications of data science to physics-based disciplines and to demonstrate the useful and game-changing applications of data science in engineering and industrial applications is to develop a new generation of engineers and scientists that are well versed in the application of data science. In other words, the objective should be to train and develop engineers who understand and are capable of efficiently apply data science to engineering problem solving.

3.2 Smart Proxy Modeling in Reservoir Engineering

Developing proxy models has a long history in the oil and gas industry and in many other industries that use numerical simulation. Proxy models provide fast approximated solutions that substitute large numerical simulation models. They serve specific useful purposes such as assisted history matching, uncertainty analysis and quantification, as well as production/injection optimization. The most common proxy models are either ROMs or response surfaces. While the former speeds up simulation run times by grossly approximating the problem, the latter accomplishes the same objective by grossly approximating the solution space. Nevertheless, they are routinely developed and used in order to generate fast solutions to changes in the input space. Regardless of the type of model simplifications used, these conventional (traditional) proxy models can only provide, at best, an approximation of the responses at the well locations, that is, of the pressure or rate profiles at the well locations.

The technology that is showcased in this book as a tool to be utilized for the analysis and effective modeling of CO$_2$ storage in geological formations is a new approach to building proxy models. One of the major advantages of this method, when compared with traditional proxy models, is its capability to replicate the results of numerical simulation models, away from the wellbores. The method is called "smart proxy modeling" because it is has the unique capability of being able to replicate the pressure and saturation distribution throughout the reservoir at the grid block level, and at each time-step, with reasonable accuracy, without sacrificing the physics or the resolution of the original numerical simulation model. Smart proxy modeling performs this task at very high speed when compared with conventional numerical simulators such as those currently in use.

Understanding the extent of changes in pressure and saturation throughout the geological formation that is used for the storage of CO$_2$, especially beyond the injection wells, is a key component in designing many reservoir engineering-related operations associated with the geological storage of CO$_2$. Examples of where it is important to have access to the dynamic behavior of pressure and saturation, which play a significant role in planning, include the following:

- Design of CO$_2$ sequestration projects where the reach of the CO$_2$ plume is an important design parameter;
- Design of CO$_2$-EOR (enhanced oil recovery) or water flooding projects; and
- Design of multi-cluster, multi-stage hydraulic fractures in shale formations to understand the extent of the stimulated volume.

Numerical reservoir simulation models are the only tools that are capable of providing design engineers with such information. Since numerical reservoir simulation models solve the flow equation at the grid block level as they simulate fluid flow throughout the reservoir, they calculate the pressure and saturation values at each time-step, for each grid block. This strength of numerical reservoir simulation models comes with a hefty price. The price to pay for this information is computational time.

Numerical reservoir simulation models solve a system of equations using iterative techniques. Convergence to a final solution usually takes several iterations. The number of iterations required for convergence usually increases as the changes in pressure and saturation become significant. This is specifically true during the transient period or when large changes in pressure and saturation may result from highly conductive media. The size of the grid blocks plays an important role in the amount of time that is required to complete a time-step, because the number of grid blocks is a function of size when a specific reservoir is being considered. The final outcome of the numerical simulation is a pressure and saturation distribution map (volume) throughout the reservoir (at each grid block) for each time-step.

In the past several decades, proxy models have become popular in the oil and gas industry. Proxy models are used to fulfill many different purposes. They are used to assist in field development planning, uncertainty analysis and quantification, optimization of operational design, and history matching. The current state of building proxy models, when it comes to representing numerical reservoir simulation models, leaves much to be desired. When successfully developed, proxy models can ultimately reproduce the results of numerical reservoir simulation models at well locations. These results are usually limited to pressure or production profiles. When it comes to changes in pressure, saturation, and molecular concentration of CO_2 throughout the reservoir (between wells), which in numerical simulation models are represented by pressure, saturation, and molecular concentration of CO_2 values at each grid block, proxy models fall short. Unless we are willing to treat each grid block in the model (which may include millions of grid blocks) as an individual well, and provide response surfaces at every grid block, an impractical and wasteful approach, no other viable solution exist, until now.

Smart proxy modeling was first introduced as the "surrogate reservoir model" (SRM) in 2006 (20–22). The original SRM only dealt with pressure and fluid rate profiles at the well, and therefore was later named "well-based SRM" (23,24). The next step in the development of smart proxy modeling was to generate smart proxy models at the grid block level. At that time this technology was called "grid-based SRM." The objective of grid-based SRM is fast reproduction of the numerical simulation model's results, changes in pressure and saturation, as a function of time and space at the grid block level, with high accuracy.

Accomplishing this objective grid-based SRM integrates reservoir engineering and reservoir modeling with machine learning and data mining. Grid-based SRM attempts to learn the mechanics of fluid flow in the porous media from the data generated by the numerical simulation model in order to reproduce it for all kinds of scenarios that it may or may not have seen during the training process. This unique characteristic of grid-based SRM makes it accurate and quick. Grid-based SRM accurately replicates the pressure, saturation, and molecular concentration of CO$_2$ distribution throughout the reservoir (at every grid block) at a very fast speed. This feature allows fast-track analysis of complex reservoir simulation models as well as the design and optimization of complex development scenarios in record time.

3.3 Mechanics of Smart Proxy Modeling

The design and development of grid-based SRM is not trivial. It requires a reasonably deep understanding of artificial intelligence, data mining, and machine learning and their application in reservoir modeling and reservoir engineering. The process of developing grid-based SRM can be summarized as using data from a numerical reservoir simulation model to teach reservoir engineering to a machine, or for now, to a computer program.

This may sound a bit unfamiliar at first, but let us take a few minutes and see what this phrase means and how it can be accomplished. We teach reservoir engineering to our future engineers using mathematics and physics, with English as the language and the means for communication. We develop equations that govern fluid flow through porous media. Using physical principles such as Darcy's law and the diffusivity equation, we explain the role of changes in pressure, permeability, and fluid viscosity in the amount of fluid that can move throughout the porous medium.

Since we cannot explain these physical phenomena to a machine (at least not yet), and since the language that machines (the original algorithm that is used for machine learning) understand is data, we have to try to explain (teach) these concepts to the computer through the use of data in a process that is called training. Machine learning works with input–output pairs of data. The above-mentioned physical principles (fluid flow through porous media) should be summarized in the input–output pairs of data that are used for training purposes.

This includes the design of the input–output pairs of data such that they carry the information content that needs to be transferred to the computer in the form of generalizable knowledge. This is the information that can be called upon for response to new situations. For the purposes of the grid-based SRM that is presented in this chapter, the data is generated from the runs made by the numerical simulation and very much depends on the specific reservoir (or flow process) that is being modeled. This technology can be generalized such that it can eventually lead to a completely new way of performing reservoir simulation and modeling.

Knowledge of reservoir engineering and reservoir modeling helps us in designing the input–output pairs of data such that they carry the required information content, while knowledge of machine learning and data mining helps us in designing the data structure and in identifying the most appropriate architecture and algorithms for training the data-driven models. The process of developing the dataset that is needed for the training process includes summarization, abstraction, and preparation.

Grid-based SRM offers a new way of utilizing numerical reservoir simulation models where changes in pressure and saturation throughout the reservoir can be incorporated into the system models, optimization routines, history matching problems, uncertainty analysis, and risk mitigation studies, all at the same resolution as the numerical simulation model. Grid-based SRM may be coupled with well-based SRM so that it can comprehensively represent all the functionalities of a numerical reservoir simulation model.

In order to successfully train a grid-based SRM the problem should be treated as a "path function" where changes in pressure and saturation are tracked both in time and in space. This is one of the major differences between smart proxy modeling and using statistics to generate "response surfaces." Response surface development as a proxy for the numerical simulation models is a traditional way of developing proxy models. Response surfaces are by definition "point functions" and that is the reason behind their lack of generality and their limited application.

As mentioned before, in the context of the smart proxy modeling, a data record that consists of an input–output pair should include all the required information about the grid block being studied (which we will refer to as the active or focal grid block) along with all the neighboring grid blocks that are in contact with it, as shown in Figures 3.1 and 3.2. All the static and

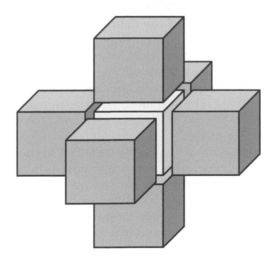

FIGURE 3.1
Active grid block with its neighboring blocks.

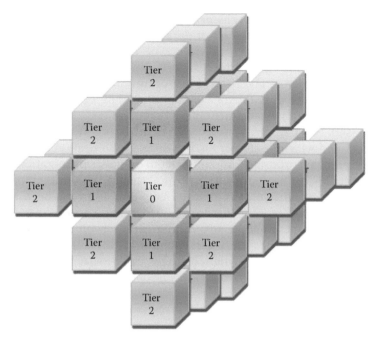

FIGURE 3.2
Every grid block in the model is used as a training record. All the neighboring grid blocks contribute to the model response of the target grid block.

dynamic information for these grid blocks must be included in each data record. The static set of information includes the active grid block's location and its relative distance from the dynamics of the reservoir such as all the inner and outer boundaries that are defined during the modeling process. This is shown in Figure 3.3.

Other static information may include reservoir characteristics for the active/focal grid block being modeled and all the connected grid blocks (Figures 3.1 and 3.2). No transmissibilities are explicitly calculated during the development of the grid-based SRM. All transmissibilities are implicitly incorporated through the use of the geometry and reservoir characteristics of all the involved grid blocks.

Dynamic information that should be included in the data compilation process includes the constraints that are imposed on all the active wells in the field, as well as the pressure, saturation, and CO$_2$ molecular concentration values of all the involved grid blocks (active/focal and neighboring blocks) at one time-step behind. The spatio-temporal dataset that is assimilated using the above information must have all that one wishes to teach an SRM.

This dataset should be analyzed carefully so that it has the right amount of redundancy as well as all the required information. It needs to be noted that

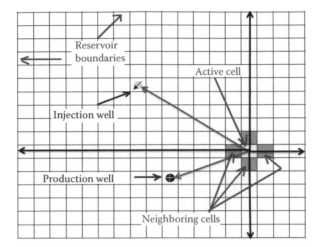

FIGURE 3.3
Location of the active block with respect to the dynamics of the reservoir, shown in two dimensions.

the sheer number of grid blocks that are included in a reservoir simulation model (sometimes multi-million grid blocks) can make this effort quite a challenging one. It is important to realize that since during the development of the grid-based SRM we try to teach reservoir engineering to a computer and that we try to accomplish this task using data, we need examples that are not the same (redundant*).

Therefore, one can see that grid-based SRM thrives on the heterogeneity of the reservoir. In other words, when the reservoir is homogenous, there is not much that we can teach the grid-based SRM and our work gets much harder. When modeling academic or toy problems, we tend to assume homogeneity in order to be able to understand the problem better, and by eliminating much of the real world's complexities we try to minimize variability during much of the study. Although such an approach has plenty of educational benefits, it makes the development of a grid-based SRM very difficult. It is strongly recommended not to test the capabilities of grid-based SRM using toy problems.

3.4 Initial Examples of Smart Proxy Modeling

As mentioned previously, smart proxy modeling was developed for the oil and gas industry, and its initial applications concentrated on oil, gas, and

---

* Redundancy is not a bad thing in machine learning, but it must not be the dominating force.

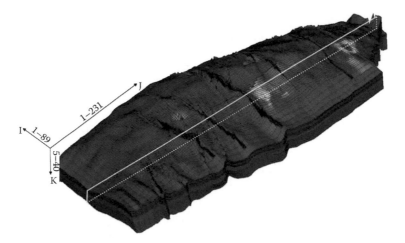

FIGURE 3.4
The full field numerical reservoir simulation model representing a giant oilfield in the Middle East.

water production in hydrocarbon-producing reservoirs. To demonstrate the capabilities of smart proxy modeling and its application, some examples from the original development of this technology are presented in this section.

3.4.1 A Giant Onshore Oilfield in the Middle East

Figure 3.4 shows the numerical reservoir simulation model that represents a giant oilfield in the Middle East. The total daily oil production from this field is capped at 250,000 barrels per day. Furthermore, each well is capped at 1,500 barrels of liquid per day. Since water is being injected as part of a pressure maintenance program, and since water cut has been a problem in some wells, the production cap is imposed in order to avoid bypassing oil and creating hard to produce oil banks that are left behind. On the other hand it was suspected that several wells in the field might be capable of producing more oil without the threat of a high water cut and a carefully planned rate relaxation program was desired and became the main objective of the project.

The operation in this field includes water injection into some of the layers for pressure maintenance and sweep purposes. Gas injection is also taking place in some areas of the field. The reservoir includes many major and minor faults that have been detected by geoscientists and are part of the geological model that has been used to build the full field numerical simulation model. Several rock types have been identified in this reservoir and have played an important role in developing the geological, and later the dynamic, model. The dynamic model has been developed using ECLIPSE™* and includes about one million

---

* Eclipse is the commercial numerical reservoir simulation model offered by Schlumberger Service Company.

grid blocks.* A single run of the version of the dynamic model used for this study took about 10 hours on a cluster of twelve 3.2 GHz Intel Xenon CPUs. This reservoir's simulation model includes more than 800,000 grid blocks.

Changes in water saturation were one of the results from this numerical simulation model that were of importance to the asset management. Since water was being injected for pressure maintenance and oil displacement, and given the fact that an increase in the water cut was being observed in several wells, it was important to be able to estimate movement of the water through the reservoir.

Uncertainties associated with the static model, which is an important issue that all numerical reservoir simulation models must deal with, complicate such estimations. Therefore, it is important to quantify the uncertainties associated with water distribution throughout the reservoir. This requires access to high-speed interactions with the model, which would be impractical given the slow nature of the numerical reservoir simulation model.

A smart proxy model was developed to assist in this effort. This was trained and validated using only eight (8) simulation runs. Upon completion of the development, which included training, calibration, and validation using

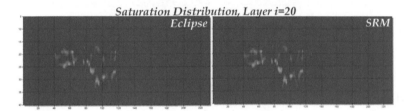

FIGURE 3.5
Comparison of the saturation distributions from a commercial numerical simulation model (left) and the smart proxy model (right) in a 2D slice for layer 20 at a given time-step.

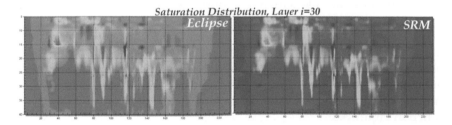

FIGURE 3.6
Comparison of the saturation distributions from a commercial numerical simulation model (left) and the smart proxy model (right) in a 2D slice for layer 30 at a given time-step.

---

* The reservoir simulation model has been developed by a national oil company (NOC). All runs were made by the asset team at the NOC and only the results of the runs were shared with those involved in the development of the smart proxy model.

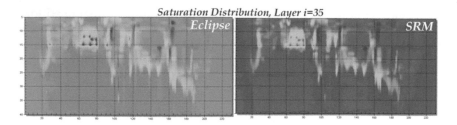

FIGURE 3.7
Comparison of the saturation distributions from a commercial numerical simulation model (left) and the smart proxy model (right) in a 2D slice for layer 35 at a given time-step

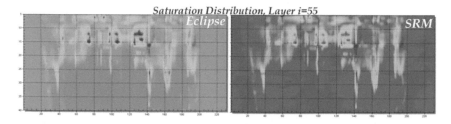

FIGURE 3.8
Comparison of the saturation distributions from a commercial numerical simulation model (left) and the smart proxy model (right) in a 2D slice for layer 55 at a given time-step.

blind simulation runs, the smart proxy model was able to accurately replicate changes in pressure and saturation throughout the reservoir at every grid block and at every time-step. The trained smart proxy model is an extremely fast tool and can be a vital tool for reservoir management. A single run of the smart proxy for this reservoir to replicate pressure and water saturation for every grid block takes less than 5 second to complete on an average laptop computer.

Figures 3.5 through 3.8 show a comparison of the water saturation distribution generated by the numerical reservoir simulation model and those generated by the smart proxy model (also called the surrogate reservoir model [SRM] as mentioned in the previous pages of this chapter) of a blind run for several 2D slices of the reservoir. The orientation of these layers is shown in Figure 3.4.

4

CO$_2$ Storage in Depleted Gas Reservoirs

Shahab D. Mohaghegh and Shohreh Amini

CONTENTS

Depleted gas reservoirs are one of the major geological formations used for the storage of CO$_2$. In this chapter we examine the application of data-driven analytics in CO$_2$ storage in a depleted gas reservoir. The depleted gas reservoir that is the subject of this chapter is in southeast Australia. The field is located in Otway Basin in Victoria, Australia (Figure 4.1). This field was identified as a suitable option for CO$_2$ sequestration because the site is well characterized due to its natural gas production history. Furthermore, having previously stored natural gas for millions of years it obviously has a proven storage capacity for the safe storage of CO$_2$ (25).

The target reservoir for CO$_2$ storage is Waarre-C Formation, which is a sandstone reservoir approximately 100 ft thick. This formation is located 6,561 ft below sea level with an area that covers about 500 acres. The reservoir is overlain by a mudstone (Flaxman and Belfast Formations) cap-rock. The structure is bound by three major sealing faults and two aquifers, which are connected to the reservoir from the southeast and west sides. The reservoir has an average porosity and permeability of 15% and 1000 mD, respectively (26).

As a depleted gas reservoir two wells had already been drilled and completed in this field. Of the two existing wells one is a producer and one an injector well. The producer well, which later was switched to an observer (monitoring) well, is named Naylor-1. This well produced natural gas from Otway Formation from May 2002 to October 2003. Data from this period of production provided valuable information that was used to history-match the numerical reservoir simulation model developed for this field. The conversion of Naylor-1 to a monitoring well took place in 2007 when the second well,

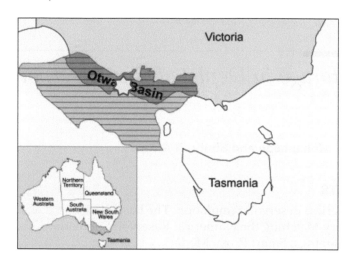

FIGURE 4.1
Otway Field location in Victoria, Australia. (From Sharma, S. et al. 2009. *Energy Procedia*, Vol. 1, pp. 1965–1972.)

CRC-1, was drilled and completed. CRC-1 was used as a CO$_2$ injection well. This well is located approximately 980 ft away from Naylor-1. During this pilot project, the intent of which was to examine the storage capabilities of the Otway Formation, CO$_2$-rich gas containing about 80% CO$_2$ and 20% methane, produced from a nearby gas field, was compressed, transported, and then injected into the Otway reservoir using the CRC-1 well.

CO$_2$ injection through CRC-1 started in March 2008 (26). Figure 4.2 is a schematic diagram that shows the location of these two wells (CRC-1 as the injection well and Naylor-1 as the monitoring well). The way CO$_2$ is transported and injected into the reservoir for long-term storage is also shown in this figure (27).

Completion of this study included three major steps: (1) development of a numerical reservoir simulation, (2) history-matching of the numerical reservoir simulation model using historical production, and (3) development of a smart proxy model to accurately represent the numerical reservoir simulation model at high speeds, as is required for all relevant studies such as quantification of uncertainties and field development planning.

4.1 Numerical Reservoir Simulation: The Base Model

The first step of this study, as already mentioned, included the development of a numerical reservoir simulation model to simulate natural gas production as well as the CO$_2$ injection processes in this reservoir. The reservoir model was

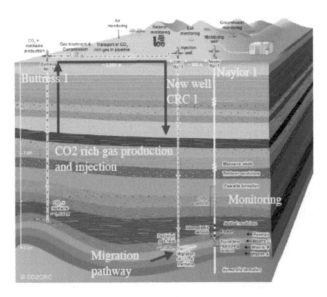

FIGURE 4.2
Schematic of the Otway CO₂ injection project. (From Sharma, S. et al. 2009. *Energy Procedia*, Vol. 1, pp. 1965–1972.)

developed in CMG-GEM (Computer Modeling module) using all available static and dynamic data from the field.

The structure of the reservoir was generated using a contour map available from the geological information of the field (Figure 4.3). The static (geological) model developed for this field included 100×100 grid blocks in the x–y

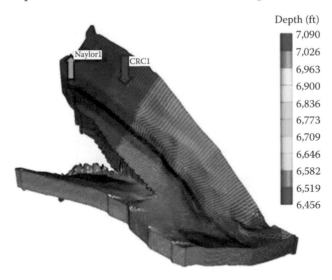

FIGURE 4.3
Reservoir model structure and well locations.

dimension and consisted of 10 simulation layers. Therefore the total number of cells (grid blocks) for the numerical reservoir simulation model ended up being $100 \times 100 \times 10 = 100{,}000$. The original production well (Naylor-1) was completed in two layers (simulation layers 5 and 6), while the injection well (CRC-1) was completed in five layers (simulation layers 3 through 7).

Available well logs and core data, examples of which are shown in Figure 4.4, were used to generate reservoir characteristics such as permeability and porosity, as required by the numerical reservoir simulation. Well logs and core analysis data provided reservoir characteristics at the well location. As is common practice in the reservoir modeling community, these values at the well location were used in conjunction with geo-statistical techniques (such as the Kriging technique) to generate reservoir characteristics for all 100,000 grid blocks in the reservoir simulation model. Permeability and

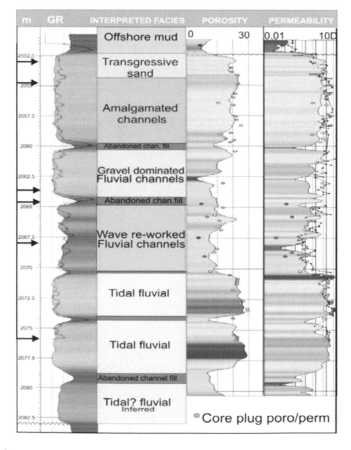

FIGURE 4.4
Porosity and permeability values for the numerical simulation model were extracted from core and log data. (From Dance, T., Spencer, L., and Xu, J. 2009. *Geological Characterization of the Otway Project Pilot Site: What a Difference a Well Makes*, s.l.: Elsevier, Vol. 1(1), pp. 2871–2878.)

porosity values calculated and measured from well logs and core analysis were correlated in order to assist the geo-statistical population of the static model. These correlations are shown in Figure 4.5.

Above and beyond static reservoir characteristics, fluid characteristics must also be identified and used in the numerical simulation model. This is true

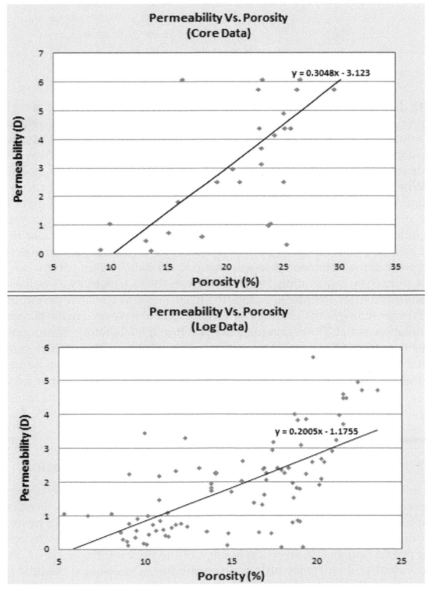

FIGURE 4.5
Correlation between porosity and permeability using log and core data.

TABLE 4.1

Composition of Natural Gas
Produced from the Otway Field

Composition	Mole Percentage
C_1	84.35
C_2	4.62
C_3	2.01
C_{4+}	1.48
N_2	6.52
CO_2	1.02

both during the development of the base model and in history matching it to represent the historical performance of the wells and the field, as well as for the fluid (CO_2) that will later be injected into the reservoir. Table 4.1 presents the composition of the natural gas produced historically from Otway Field. This composition was incorporated into the numerical simulation model.

When multi-phase flow in the porous media is being modeled using the numerical simulation, relative permeability curves are an essential input into the model. Furthermore, when the multi-phase flow in the porous media includes both production and injection, as is the case in this particular study, the hysteresis phenomenon plays a significant role. In the early injection period, drainage is the dominant process; however, imbibition takes place when the CO_2 plume starts migrating. Hysteresis affects the CO_2 mobility as well as the gas water contact. In order to include the hysteresis effect, relative permeability curves (including both CO_2 drainage and imbibition) were used in the numerical simulation model. The data were generated through laboratory measurements on a core from CRC-1 well. Due to the existence of an active aquifer system, a fraction of the injected CO_2 was dissolved into the water phase, so in the numerical simulation model, CO_2 solubility in water was also taken into account.

4.2 History Matching the Numerical Reservoir Simulation

In numerical reservoir simulation modeling, history matching is a process through which the static (and sometimes the dynamic) characteristics of the reservoir model are modified until they match the observed, historical production/injection from the field. If a reservoir model fails to match the production and/or injection history of the field (all the wells involved, individually), it will enjoy little or no credibility in forecasting the fluid flow behavior in that reservoir into the future.

Therefore, when it comes to reservoir models, specifically when CO_2 injection is a consideration, history matching is a necessary step. This stems from the fact that reservoir models do not offer a unique solution. One of the major reasons for the rise of stochastic reservoir modeling is this non-uniqueness of reservoir models. What is achieved by history matching is the internal consistency* of the model. In order to develop more confidence in a numerical reservoir simulation model's results and then its forecast into the future, having an acceptable history match of the simulation model cannot be over-emphasized.

To history match the base numerical simulation model that was developed for Otway CO_2 injection project, actual observations and field measurements from the Otway Field were required. The available field measurements included production and injection measurements since the drilling and completion of the first production well in the field. As mentioned already, well Naylor-1 was drilled, completed, and put on production in 2002, and produced for 18 months, producing natural gas from the Otway Field while well CRC-1 was drilled, completed, and used for CO_2 injection in 2008. Therefore, for history matching purposes, 18 months gas production from well Naylor-1 as well as 8 months of CO_2 injection from well CRC-1 were available.

During the history matching process, bottom-hole pressure values (18 months of gas production and 8 months of CO_2 injection) from both wells were matched. This task was achieved by incorporating modifications to the porosity and permeability of the reservoir, while the thicknesses of the reservoir layers were kept constant and not subject to modification to achieve a history match. The results of the history match of CO_2 injection from the Otway numerical reservoir simulation model are shown in Figure 4.6. In this figure, the bottom-hole reservoir pressures at two locations in the field (at the production [Naylor-1] and injection [CRC-1] wells) are superimposed and displayed. Blue and red dots in Figure 4.6 show actual field measurements, while the continuous blue line represents the bottom-hole reservoir pressure of the producer [Naylor-1] generated by the Otway numerical reservoir simulation model. The x axis in this figure shows time from May 2002 when the production well started producing until mid-2008 when CO_2 injection began in this field.

As can be observed from this figure, the numerical reservoir simulation is capable of accurately history matching the observed production and injection in this reservoir during the time of production from Naylor-1 as well as during CO_2 injection several years after gas production was abandoned and the reservoir pressure had a chance to increase. This is an important achievement, as it allows the team to move forward with the next steps of analysis and to be able to use this numerical simulation model for further analysis of CO_2 injection in the reservoir.

---

* Definition for "Internal consistency": In statistics and research, internal consistency is typically a measure based on the correlations between different items on the same test. It measures whether several items that propose to measure the same general construct produce similar scores.

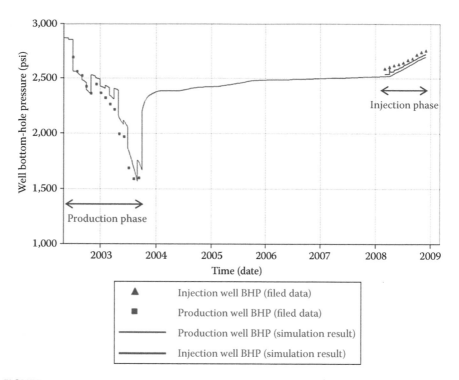

FIGURE 4.6
History match of the bottom-hole pressure (BHP) for both the production well and the injection well.

Once the history matching is successfully completed, as one of the major steps that is required during any modeling and analysis of the geological storage of CO$_2$, the model can be used for further analyses. It must be noted that although a history-matched numerical reservoir simulation model is a necessity for the detailed study and analysis of any CO$_2$ injection project in a depleted hydrocarbon reservoir, it is by no means a sufficient tool. This stems from the fact that numerical reservoir simulation models are uncertain by nature and, as such, require extensive uncertainty analysis and uncertainty quantification to be performed anytime they are to be used as the basis for decision making or planning.

Uncertainty analysis and uncertainty quantification comprise a process that requires the objective function (in this case the numerical reservoir simulation model) to execute (simulation runs) hundreds or thousands of times, any time that it is used to perform a specific analysis, in order to provide the required information for decision making. Numerical reservoir simulations used for realistic large hydrocarbon reservoirs usually include tens of millions of grid blocks (cells) and are well known to have a very extensive computational footprint. Most major oil and gas companies use hundreds of parallel central processing units (CPUs) to run their numerical reservoir simulation models. Even in such cases when hundreds of parallel CPUs are used, a single execution (run) of a large and realistic numerical reservoir simulation model takes a large number of hours. This has a lot to do with the reality that often a massive number of geological realizations need to be examined and investigated during any realistic uncertainty analysis and uncertainty quantification study.

4.3 Developing a Smart Proxy Model

The philosophy behind smart proxy modeling and its characteristics and differences when compared with traditional proxy modeling in the numerical reservoir simulation community (reduced order models and response surfaces) were discussed in Chapter 3 and therefore will not be repeated here. In this section we summarize the development of smart proxy modeling for the Otway project and present the results that have been achieved. The smart proxy in this project is referred to as the surrogate reservoir model (SRM).

The first step in the development of a proxy model is the design of its structure. The structure of the smart proxy model is mainly a function of the final objective of the project. All other activities for the development of a smart proxy will stem from this specific step that defines the final product which is usually determined by the project objective. For example, for the Otway project, since uncertainty analysis and quantification regarding the degree of migration of CO_2 into the formation as a function of CO_2 injection rates were the objective of the project, the smart proxy must be able to accurately map the reservoir pressure and the fluid saturation throughout the reservoir. Therefore, the final product of this smart proxy must be able to reproduce the reservoir pressure and the fluid saturation at every grid block in the numerical reservoir simulation model with high accuracy and of course at a fraction of the time that it takes to run the actual numerical reservoir simulation model. For this purpose, the workflow demonstrated in Figure 4.7 was developed to generate and use the smart proxy model for this project.

As shown in Figure 4.7, the workflow for the development of the smart proxy for the Otway project includes four major steps. The first step, as covered in the previous two sections of this chapter, includes building a numerical reservoir simulation model and history matching it using the available dynamic data from the field. The rest of the workflow will use this history-matched numerical reservoir simulation model for its purposes. The next step in the workflow is to collect the required amount of data from the numerical reservoir simulation model and to generate a database to be used for the development of the smart proxy model. To accomplish this objective we have to identify the number of required simulation runs that can generate the amount of data needed to train, calibrate, and validate the set of data-driven

FIGURE 4.7
Workflow for the development of the smart proxy model for the Otway Project.

predictive models (neural networks) that will form the main engines of the smart proxy for this project.

Once the required simulation runs are identified, the numerical reservoir simulation model will be executed for the identified runs, and the required data is extracted from the simulation runs and a spatio-temporal database is generated as the basis for the smart proxy model. This database is then used to train, calibrate, and validate the set of neural networks for the smart proxy model. The next step is the development of the main engines of the smart proxy model, and to assemble them together to form the final product. This includes the training, calibration, and validation of all neural networks that have been identified during the initial design of the structure of the smart proxy model.

4.3.1 Design of the Simulation Runs

The Otway project presented in this chapter was actually part of a larger project. The main project included a system model with multiple components. The system model's intention was to examine the impact of different scenarios on the overall objectives of CO_2 injection into this depleted gas reservoir. Components of the main projects included a pipeline delivery (of CO_2) model, a subsurface model (the numerical reservoir simulation model, i.e., the subject of this chapter), a wellbore model, a surface facility model, and an economical calculation model. In order for the system model to perform efficiently, all components must work interactively and must have a small computational footprint. Since the subsurface model in situations such as this requires a numerical reservoir simulation model in order to make the overall performance of the system model relevant (accurate), development of a proxy model that can accurately reproduce the results of the subsurface model (numerical reservoir simulation model) in a very short period of time becomes an essential step in the overall project.

To design the required numerical reservoir simulation runs to be used in the development of the smart proxy model, two parameters were considered as variables to be studied in detail. These variables were (1) the total amount of injected CO_2 and (2) the length of time it takes to inject this amount of CO_2 into the formation. The actual amount of CO_2 historically injected into the formation in the Otway project was 593 MMscf. This amount was injected in a period of 8 months. These values were identified as the starting point of this study. The maximum amount of injected CO_2 to be investigated was identified to be three times the original amount, and the maximum length of injection time considered was five times the original period. Ten simulation runs were designed within these limits, as shown in Figure 4.8. The actual values for the selected scenarios are shown in Figure 4.9. Based on the numbers shown in Figure 4.9, ten different injection schedules were designed to be used in the numerical reservoir simulation runs. Details of these designed schedules are shown in Figure 4.10.

Once the numerical reservoir simulation runs were designed, the next step was to execute these runs and extract all the required data from each of the runs. These extracted data are used to build the spatio-temporal database to be used as the foundation of the smart proxy model.

4.3.2 Spatio-Temporal Database

The numerical reservoir simulation model in this study was developed using Computer Modeling Group's* (CMG) reservoir simulator. The model

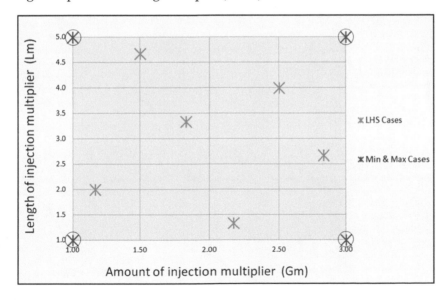

FIGURE 4.8
Selection of the simulation runs to be used in the development of the smart proxy model.
LHS: Latin Hypercube Samples.

---

* http://www.cmgl.ca/

G (MMScf)	L (Month)
1.68E+03	21
8.90E+02	37
1.48E+03	32
6.92E+02	16
1.29E+03	11
1.09E+03	27
5.93E+02	8
1.78E+03	8
5.93E+02	40
1.78E+03	40

FIGURE 4.9
Calculated amount and length of injection time for the 10 simulation scenarios.

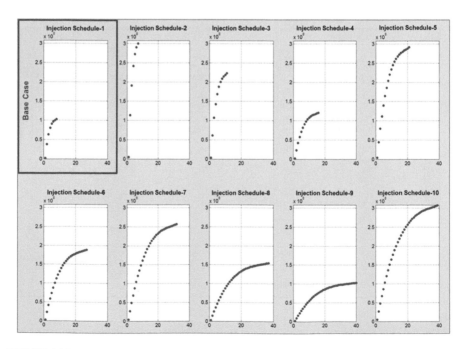

FIGURE 4.10
Injection schedule generated for the 10 simulation scenarios.

included 100,000 (100 × 100 × 10) grid blocks. Therefore, the ten simulation runs that were designed and executed for the development of the smart proxy generate a very large number of records. To put this in perspective, it should be mentioned that each simulation run will generate 100,000 records for each time-step. Given the fact that we have a total of 240 time-steps, the total number of records in the spatio-temporal database will add up to 24,000,000 records. Each record in the spatio-temporal database represents the static and dynamic characteristics of a single grid block in a given run and given time-step. The list of collected static and dynamic parameters for each grid block is shown in Figure 4.11.

The spatio-temporal database generated during this step of the process must include all the information that needs to be used to train the set of neural networks that form the smart proxy model. Therefore, the spatio-temporal database must be assimilated in the context of the physics of the problem at hand. Since by performing this data-driven predictive modeling we are essentially attempting to teach the smart proxy reservoir engineering, we must assimilate the spatio-temporal database in the context of the fluid flow in the porous media. This means that each record in the spatio-temporal database must include the relevant information on how pressure, fluid

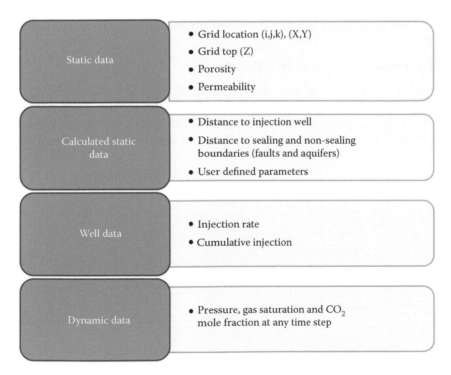

FIGURE 4.11
List of static and dynamic data used for the generation of the spatio-temporal database.

saturation, and CO_2 mole fraction change as a function of time and space in the reservoir simulation model. In order to accommodate this most important item in the development of the smart proxy, a tier system of grid blocks in relationships with each single grid block is considered. It is a fact that when CO_2 is injected into the reservoir, it has the potential to move in all directions based on the existing static and dynamic characteristics of the reservoir at any given time-step. Therefore, changes in reservoir pressure, CO_2 mole fraction, and gas saturation become a function of the characteristics (such as porosity and permeability) of the lower and upper layers for each grid block. Also, it is not just a single grid block in the upper and lower layer that impacts the distribution of dynamic parameters; rather a group of grid blocks are contributing to this process.

Therefore, in order to take into account the impact of this interdependency, a new scheme is used for the tier system calculation, as shown in Figure 4.12. To obtain the values for Tier-1, Tier-2, and Tier-3, the average values of the parameters are calculated over the grids that are included in the corresponding tier system. It should be mentioned that the number of grid blocks included in each tier system can be increased (or decreased) in the case where the size of the reservoir grid blocks is modified. Therefore, information about each tier system for every grid block in each time-step is also included in the spatio-temporal database. While this effort makes no change to the number of records in the spatio-temporal database, it does increase its dimensionality.

4.3.3 Data-Driven Predictive Models

Once the generation of the spatio-temporal database is completed, it is used as the main source of data and information for the development of the data-driven predictive models that form the main engines of the smart proxy model. In this study the data-driven model of choice was the artificial

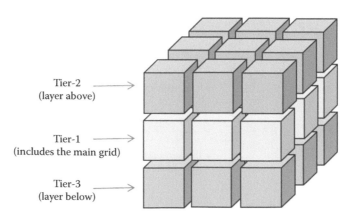

Tier-2
(layer above)

Tier-1
(includes the main grid)

Tier-3
(layer below)

FIGURE 4.12
The grid block tier system used for development of the spatio-temporal database.

neural network. Details about artificial neural networks have been covered in Chapter 2 and therefore will not be repeated here. However, we need to discuss the training, calibration, and validation scheme that was used. It must be mentioned that only a portion of the data that was generated from the ten simulation runs covered earlier in this chapter is used to develop the data-driven models. The idea is to select a portion of the data that has sufficient information regarding the range of changes in reservoir pressure, CO_2 mole faction, and gas saturation that take place throughout the reservoir and in all the time-steps involved. For this project we were able to select about 10% of the spatio-temporal database for this purpose.

The selected data is divided into the three segments of training, calibration, and validation. The neural network models are generated for all three individual dynamic parameters (reservoir pressure, CO_2 mole faction, and gas saturation). Of course, the calibration and validation datasets (together they form 20% to 30% of the data used for development of the neural network models) are considered to be blind data since they do not play any role in modification of the synaptic connection between neurons during the training process. However, in the context of development of the smart proxy for reservoir modeling, we do not consider the predictive capabilities of the developed neural network on the portion of data in the calibration and validation datasets as sufficient indicators of the robustness of the smart proxy models. For these portions of the data we usually achieve very high accuracy (R^2 values higher than 0.95). For this reason we have designed a new way of examining the predictive capabilities of the smart proxy model and its robustness. We call this examination of the smart proxy's predictive capabilities and robustness the "blind validation process."

The blind validation process includes making a completely new simulation run with characteristics that have not been used during construction and training of the smart proxy. In the case of the Otway project, this means designing and executing an 11th simulation run (we have already made ten runs for construction of the smart proxy model) to be used as the blind run and to be used as a yardstick in the evaluation of the predictive capabilities and robustness of the developed smart proxy model. Therefore, the results that will be presented in the next section include both the results from the smart proxy on the training, calibration, and validation database (the original ten simulation runs) as well as the results generated from applying the smart proxy to the blind (the 11th) simulation run.

4.4 Results and Discussions

As a reminder, the objective of this project was to develop a smart proxy model for the numerical reservoir simulation model of the Otway CO_2

geological storage project that is capable of reproducing the reservoir pressure, CO_2 mole fraction, and gas saturation values at every grid block and every time-step with high accuracy and at a fraction of the run time of the numerical simulation. The results presented in this section are divided into two parts. Part one includes the results of the development of the smart proxy. The development process includes training, calibration, and validation of the model using the data generated by the ten simulation runs that have been assimilated into the spatio-temporal database. Furthermore, it needs to be mentioned that the smart proxy was developed using 10% of the collected data from which 80% was used for training and the other 20% for calibration and validation.

Results that represent the development process of the smart proxy for the Otway project are shown in Figures 4.13 through 4.16. In each of the four figures that show the results of the smart proxy development (training, calibration, and validation), there are nine graphs grouped in three rows that include three graphs per row. Each figure includes the results of the smart proxy model (identified in the figure as SRM) compared with the results generated from the numerical reservoir simulation model (identified in the figure as CMG). The third graph in each row shows the difference (error) between the two graphs. Each figure shows a sample of a single layer

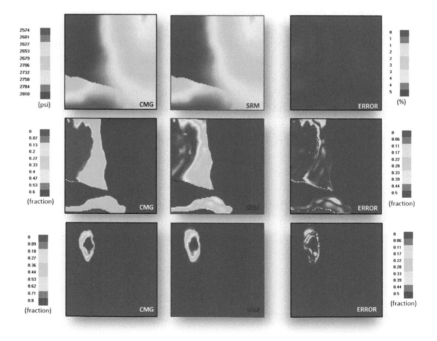

FIGURE 4.13
Distribution maps of pressure (top), gas saturation (middle), and CO_2 mole fraction (bottom): Development – run number 3; Layer: one; Time-step: month 8.

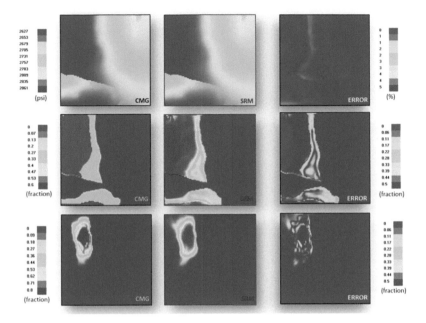

FIGURE 4.14
Distribution maps of pressure (top), gas saturation (middle), and CO$_2$ mole fraction (bottom):
Development – run number 3; Layer: one; Time-step: month 16.

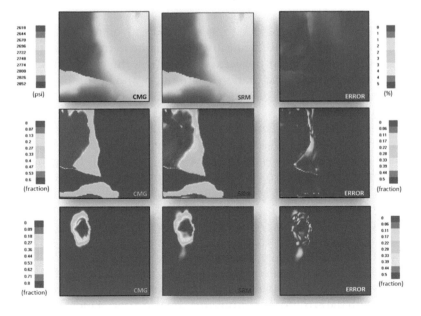

FIGURE 4.15
Distribution maps of pressure (top), gas saturation (middle), and CO$_2$ mole fraction (bottom):
Development – run number 7; Layer: two; Time-step: month 4.

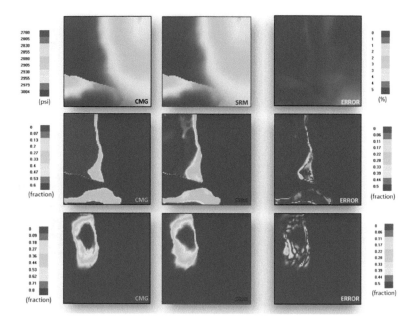

FIGURE 4.16
Distribution maps of pressure (top), gas saturation (middle), and CO$_2$ mole fraction (bottom): Development – run number 7; Layer: two; Time-step: month 8.

(100 × 100 grid block) for a given time-step and a given run that was used to build the smart proxy. In all the figures the graphs in the first row represent the results for the reservoir pressure distribution, the graphs in the second row the results for the gas saturation distribution, and the graphs in the third row the results for the CO$_2$ mole fraction distribution.

The blind (11th) simulation run that was designed to validate the predictive capabilities and robustness of the smart proxy model has characteristics that differ from the runs with characteristics displayed in Figures 4.8 through 4.10. The blind run has a total injection volume twice as much as the actual history match simulation run. Figures 4.17 and 4.18 present comparisons between the smart proxy results and the results from the numerical simulation run for two samples of reservoir pressure, gas saturation, and CO$_2$ mole fraction distributions.

The two-dimensional distribution maps presented in Figure 4.13 through 4.18 assist in visualizing the quality of the results generated by the smart proxy and the numerical reservoir simulation model. In general, the results from the smart proxy for the reservoir pressure distribution have the highest accuracy, with the error for the blind run scenario after 24 months of injection being about 2%.

For the prediction of gas saturation (for the same blind case after 24 months of injection), 85% of the results generated by the smart proxy have an error of

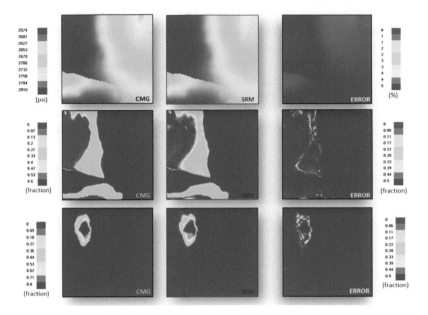

FIGURE 4.17
Distribution maps of pressure (top), gas saturation (middle), and CO_2 mole fraction (bottom): Blind run – Layer: one; Time-step: month 4.

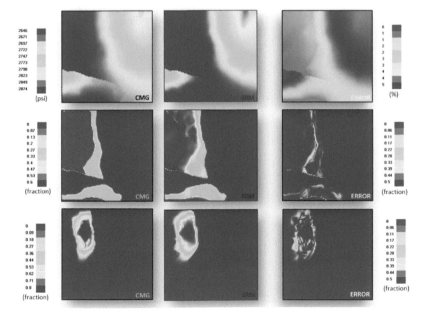

FIGURE 4.18
Distribution maps of pressure (top), gas saturation (middle), and CO_2 mole fraction (bottom): Blind run – Layer: one; Time-step: month 8.

less than 0.02 (less than 2%), and for the CO$_2$ mole fraction 95% of the results generated by the smart proxy have an error of less than 0.05 (less than 5%). In general, the smart proxy can be trusted to reproduce the dynamic parameters calculated by the numerical reservoir simulation model with less than 10% error, but in a fraction of the time. The smart proxy runs are calculated in seconds (and sometimes in few minutes) with no need for clusters of CPUs or graphics processing units (GPUs). All analyses and results presented in this chapter were performed on an average desktop computer with a single CPU.

5

CO₂ Storage in Saline Aquifers

Shahab D. Mohaghegh and Alireza Haghighat

CONTENTS

Geological formations composed of deep sedimentary rocks that are saturated with water and brine containing a considerable salt concentration are known as saline aquifers. While it is not economically or conventionally viable to use high-salinity brines for drinking or irrigation, deep saline aquifers have been used for low-power geothermal energy generation as well as for the injection of chemical waste, drilling slurries, and radioactive waste (28).

Supporting evidence suggests deep saline aquifers have enough volumetric capacity to sequester enormous amounts of CO_2. Unlike the limited available locations of oil and gas reservoirs, deep saline formations are widely spread geographically, providing more available options to store CO_2 from emission sources. Based on Yamasaki's study (29), storage capacity of saline aquifers is much more than the capacity found in oil/gas fields and un-mineable coal beds (Table 5.1). The retention time for CO_2 stored into saline aquifers is estimated to support up to thousands of years, representing the most viable storage option amongst the other geological formations. Reservoir characterization is a concern for saline aquifer storages as it drastically lacks available information in comparison with depleted oil and gas reservoirs.

5.1 Storage Capacity

Calculations of the storage capacity require estimations of total affected space which represents the whole region that is impacted by the CO_2 injection. The void space required to store injected CO_2 is created by compressing the fluids in the formation and, as a result, increasing the reservoir pressure. The ultimate storage capacity is also determined by the maximum average reservoir pressure allowed (so that it will not fracture – break – the formation rock into which it is being injected), which varies depending on the applied regulation (i.e., 10 bars or 10% of the initial reservoir pressure).

TABLE 5.1

CO_2 Storage Capacity

Option	Capacity (Gt-C)
Depleted oil fields	120
Depleted gas fields	188
Un-mineable coal-beds	11
Saline aquifers	109–2,727
Oceans	1,400–200,000

Source: Yamasaki, A. 2003. An Overview of CO$_2$ Mitigation Options for Global Warming – Emphasizing CO$_2$ Sequestration Options. *Journal of Chemical Engineering of Japan*, Vol. 36, 361–375.

In other words, most regulatory guidelines only allow enough CO_2 to be injected into a given saline aquifer such that it increases the native reservoir pressure up to 10% of the original formation pressure. Assuming rock/brine compressibility effects and the maximum allowable average pressure in a multi-layer reservoir, the maximum storage capacity can be calculated by the following formula (30):

$$V_{CO_2} = \sum_{l=1}^{L}\sum_{i=1}^{I} V_l \phi_l S_i^l (C_l + C_i) \Delta \overline{P_l}$$

where
V_{CO_2} = theoretical maximum storage capacity (reservoir m³ of CO_2)
V = bulk volume (reservoir m³)
ϕ = porosity (fraction)
S = saturation (fraction)
C = compressibility (1/Pa)
ΔP = average pressure difference $p-p_0$ (Pa)

Given the above equation, to predict the storage capacity it is necessary to obtain reasonable values for the affected space or reservoir boundary, compressibility, and maximum allowable pressure. Typically, there are three types of pressure increase that take place during the CO_2 injection process. These are:

- Local pressure increase (bottom hole pressure) – near and around the wellbore where injection is taking place
- Regional pressure increase (reservoir pressure)
- Total pressure increase (affected space pressure)

Due to the estimation techniques proposed by entities such as the Carbon Sequestration Leadership Forum (CSLF) and the United States Department of Energy, it is important to note that there are a variety of methods that may be used to calculate storage capacities in un-mineable coal-bed methane formations, oil/gas reservoirs, and saline aquifers (31).

The storage efficiency factor is another parameter in storage capacity calculations and can be determined in a manner similar to an "original oil in place" calculation. The "available space" that is covered by sealing with the cap-rock includes the total pore space for CO_2 storage. A portion of the "available space" is filled by CO_2 after completion of the injection, a function of the reservoir characteristics, and is defined as the "used space." The efficiency factor simply represents the "used space" to "available space" ratio (Figure 5.1) (30).

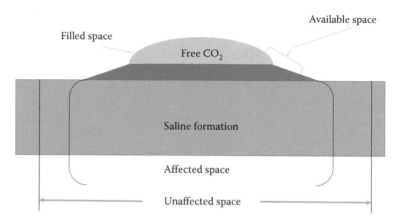

FIGURE 5.1
Different regions in a saline aquifer after CO$_2$ injection. (From Meer, L. and Egberts, P. 2008. A General Method for Calculating Subsurface CO$_2$ Storage Capacity. *Offshore Technology Conference*, Houston, TX, 241134. s.n., 2008.)

5.2 Saline Aquifer Distribution

Various studies have proposed multiple sedimentary regions all around the world that have been identified as being suitable for CO$_2$ storage. Sediments that are located in the mid-continent or close to the edge of continental plates are considered to be suitable for CO$_2$ storage due to their stability and structure. Basins located behind mountain ranges (formed by plate collision) such as the Andean mountains in South America, the Appalachian and Rocky mountains in the United States, the Alps and Carpathians in Europe, and Zagros and Himalayas in Asia are good potential locations for storage (32). Plate edges are not ideal locations for the basins due to the subduction occurring between active mountains, which creates highly folded/faulted regions and includes paths for leakage. Other important characteristics to determine good reservoir options for storage include depth, temperature gradient (colder basins are more suitable for storage), reservoir pressure, porosity, thickness, and reservoir dip.

As mentioned in the previous section, saline aquifers provide very large volumes for CO$_2$ sequestration. Based on the source–sink (coal-fired power plants–saline aquifer) distribution atlas (33), in the United States, it is estimated that more than 95% of the main CO$_2$ sources are within 80 km of a potential underground storage. Figure 5.2 shows estimates (34) for saline aquifer storage capacities in the United States for each state/province. The national total storage capacity for saline aquifers is reported to range from 1,820 to more than 22,260 million tons (low and high estimates). Texas, Louisiana, Montana, Wyoming, Mississippi, New Mexico, Colorado, California, and Washington represent the largest CO$_2$ storage resources.

*CO$_2$ Storage Resource Estimates for Saline Formations by State/Province**

State/Province	Million Metric Tons		Million Tons		State/Province	Million Metric Tons		Million Tons	
	Low Estimate	High Estimate	Low Estimate	High Estimate		Low Estimate	High Estimate	Low Estimate	High Estimate
Alabama	11,760	161,630	12,963	178,167	New Brunswick				
Alaska					New Hampshire				
Alberta	35,150	39,240	38,746	43,255	New Jersey	0	0	0	0
Arizona	120	1,580	132	1,742	New Mexico	32,120	441,650	35,406	486,836
Arkansas	4,320	59,420	4,762	65,499	New York	1,700	6,820	1,874	7,518
British Columbia	1,590	2,120	1,753	2,337	Newfoundland & Labrador				
California	30,070	413,490	33,147	455,795	North Carolina	1,320	18,170	1,455	20,029
Colorado	30,860	424,330	34,017	467,744	North Dakota	103,220	120,070	113,781	132,355
Connecticut	0	0	0	0	Northwest Territories				
Delaware	20	80	22	88	Nova Scotia				
District of Columbia	0	0	0	0	Ohio	3,970	15,900	4,376	17,527
Florida	15,750	216,910	17,361	239,102	Oklahoma	0	0	0	0
Georgia	490	23,200	540	25,574	Ontario	10	20	11	22
Hawaii					Oregon	7,080	97,390	7,804	107,354
Idaho	50	720	55	794	Pennsylvania	5,900	27,620	7,606	30,446
Illinois	8,490	115,330	9,359	127,130	Quebec	0	0	0	0
Indiana	14,370	85,440	15,840	94,181	Rhode Island	0	0	0	0
Iowa	10	150	11	165	Saskatchewan	980	8,820	1,080	9,722
Kansas	1,190	16,400	1,312	18,078	South Carolina	200	9,660	220	10,648
Kentucky	1,350	9,450	1,488	10,417	South Dakota	17,390	155,990	19,169	171,950
Louisiana	149,360	2,053,760	164,641	2,263,883	Tennessee	490	6,650	540	7,330
Maine					Texas	333,400	4,584,250	367,511	5,053,271
Manitoba	310	310	342	342	Utah	20,990	288,680	23,138	318,215
Maryland	860	5,050	948	5,567	Vermont	0	0	0	0
Massachusetts	0	0	0	0	Virginia	80	390	88	430
Michigan	14,620	58,490	16,116	64,474	Washington	29,930	411,570	32,992	453,678
Minnesota					West Virginia	4,480	17,930	4,938	19,764
Mississippi	45,450	624,940	50,100	688,878	Wisconsin	0	0	0	0
Missouri	20	310	22	342	Wyoming	87,430	1,202,200	96,375	1,325,199
Montana	120,710	1,653,720	133,060	1,822,914	Offshore	491,080	6,756,360	541,323	7,447,612
Nebraska	22,860	76,840	25,199	84,702	**TOTAL**	1,652,550	20,213,050	1,821,625	22,281,074
Nevada	0	0	0	0					

* States/Provinces with a "zero" value represent estimates of minimal CO$_2$ storage resource, while states/provinces with a blank represent areas that have not yet been assessed by the RCSPs.

FIGURE 5.2

CO$_2$ storage resource estimates for saline formations in the USA. (From National Energy Technology Laboratory, NETL. 2010. *Carbon Sequestration Atlas of the United States and Canada (Atlas III)*. Morgantown, WV: US Department of Energy.)

5.3 CO$_2$ Trapping Mechanisms

Sequestration of CO$_2$ in the saline aquifer may occur by various types of trapping mechanism. Trapping mechanisms control the movement of the injected gas in the reservoirs. CO$_2$ trapping mechanisms in saline aquifers include hydrodynamic trapping, trapping by solubility, residual trapping, and mineral trapping. In hydrodynamic trapping (also known as structural or geologic trapping), the injected CO$_2$ compresses the water/rock and occupies the free pore space of the reservoir rock. Although the compressibility of water is small, the large volumes of water and sufficient injection pressure make it possible for gas bubbles to form (30). In solubility trapping, injected gas dissolves in the aquifer based on water salinity, temperature, and pressure (Figure 5.3). Notably, complications may arise where CO$_2$ reacts with water, yielding carbonic acids or other carbonates. If the CO$_2$ causes rich brine to flow, dissolved CO$_2$ may move in the reservoir. When the brine is completely saturated with CO$_2$, an increase in water density occurs, and this phenomenon leads to natural convectional flow in the reservoir, which can enhance the diffusion rate of CO$_2$ in the reservoir brine.

The residual trapping mechanism works as CO$_2$ saturation in the reservoir reaches values below the minimum gas saturation required to initiate the flow of gas, also known as the "residual gas saturation." At this stage, the gas becomes immobile in the pore space. Residual gas saturation mainly depends on the end point relative permeability. Although residual CO$_2$

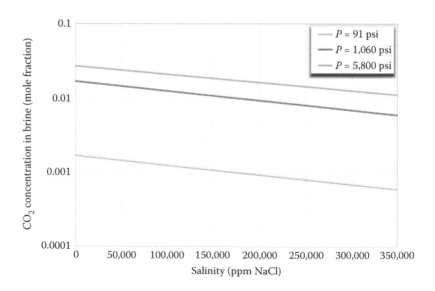

FIGURE 5.3

Effect of salinity and pressure on CO$_2$ solubility in brine (SI conversion: 1 psi = 6.9 kPa). (From Cooper, C. 2009. *A Technical Basis for Carbon Dioxide Storage.* s.l.: CO$_2$ Capture Project (35).)

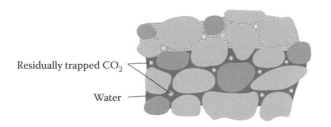

Residually trapped CO_2

Water

FIGURE 5.4
Schematic of residual trapping. (From CO2CRC. Cooperative Research Centre for Greenhouse Gas Technologies. [Online] Australian Government's Cooperative Research Centres program. http://www.co2crc.com.au.)

remains buoyant, it represents no mobility due to capillary forces as it is not connected to mobile CO_2 clusters (Figure 5.4).

The mineral trapping process occurs as dissolved CO_2 reacts with some of the reservoir rock minerals rich in calcium, magnesium, and iron and forms carbonate compounds. There is supporting evidence that for most reservoirs the mineral trapping mechanism will have a minimal impact in the first years (35). Due to an insufficient understanding of the subsurface characteristics, there are noticeable uncertainties associated with CO_2 reactions and corresponding rates.

Notably, the trapped CO_2 in the hydrodynamic mechanism represents higher potential for leakage due to the mobility of the free CO_2. In solubility, residual, and mineral trapping, the form of the geological CO_2 storage is more stable or permanent. The amount of CO_2 trapped by each mechanism at a given site will have a significant impact on site security. The focus in this chapter is on hydrodynamically trapped CO_2.

5.4 CO_2 Storage in Saline Aquifers: Case Study

In this section, practical examples of CO_2 sequestration in saline aquifer projects will be discussed briefly. These projects are located in Norway and the United States.

5.4.1 Sleipner (Norway)

Sleipner is a gas reservoir (divided into Sleipner West and Sleipner East) located in the North Sea about 250 km from Stavanger, Norway. The field, which is operated by Statoil, produces natural gas (35,880 m³/day), which contains about 9.5% of CO_2 and condensate. The producing formation is sandstone, which is located 2,500 m below sea level. Due to the sales regulations that enforced the operator to limit the CO_2 fraction to a maximum of 2.5%, separation units were installed on the offshore platform, or Sleipner T treatment platform. Since release of captured CO_2 in the atmosphere was not environmentally allowable, CO_2 sequestration into a saline aquifer of Utsria Formation was planned for this field. This was the first CO_2 storage project in the world. Utsria Formation, the target zone for storage, consists of fine-grained and high permeable sand located 800 m below sea level, with reservoir thickness ranging from 150 to 250 m. Injection started from 1996 at an approximate rate of 0.9 MMt/year. The cumulative injected CO_2 so far is 14 MMt (the planned value is 17 MMt). 4D seismic studies have indicated no CO_2 migration from the target layer into the other zones (39).

5.4.2 Snohvit (Norway)

Snohvit gas field is located in the Barnet Sea, and was developed with no surface installations. At a depth of 250–345 m below sea level, subsea production facilities were installed. The final production came from nine wells and was transported to land via a 143 km pipeline (Figure 5.5). Gas production started in 2007 with an average yearly rate of 28×10^6 m³ associated CO_2, which was removed on land and transferred back to the field to be injected into the Tubaen sandstone, which is 45–75 m thick and located 2,600 m below sea level. Injection started from 2008 and is planned to reach 31–40 Mt (40).

5.4.3 The Mississippi Test Site (US)

The objective of this project was to verify safe geological storage of CO_2 captured from coal-fired power plants in Lower Tuscaloosa Massive Sand

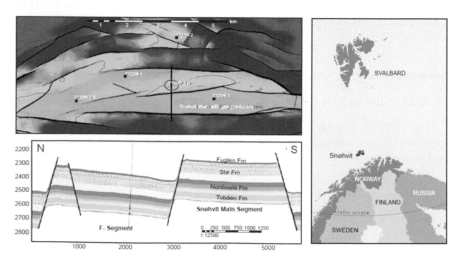

FIGURE 5.5
Location of the Snohvit fields. (From Simmenes, T. et al. 2013. Importance of Pressure Management in CO_2 Storage. *Offshore Technology Conference*, Houston, TX.)

Unit, which is located in the Gulf Coast region. Initial studies indicated that Tuscaloosa Formation may represent favorable capacity to store CO_2. This formation is located at a depth of 2,606 m and appears to have enough thickness (46–76 m), porosity (15%–33%), and extent (Figure 5.6). The Middle Tuscaloosa shale, with thickness of about 122 m, provides the primary confining unit (seal) for the target formation. An observation well, located 53 m from the injection well, was drilled into the same formation before the start of injection. In October 2008, about 3,020 tons were injected into the reservoir (the planned goal was 3,000 t of CO_2). Reservoir simulation results showed that the maximum CO_2 plume extent would be 58 m from the injection well (41).

5.4.4 In Salah (Algeria)

In Salah CO_2 storage project started in 2004 in Algeria and is operated by BP, Sonatrach, and Statoil Hydro. The reason for initiation of this project was the high concentration of CO_2 (5%–10%) in the gas produced from Krechba, Teg, and Reg fields. Export gas should contain 0.3% of CO_2. Joint venture companies spent more than $100 million to install CO_2 capture and

Stratigraphic Unit	Sub-Units	Hydrology
Misc. Miocene units	Pascagoula Fm.	Freshwater aquifers
	Hattiesburg Fm.	
	Catahoula Fm.	
Vicksburg		Saline reservoir
	Red bluff Fm.	Minor confining unit
Jackson		Saline reservoir
Caliborne		Saline reservoir
Wilcox		Saline reservoir
Midway shale		Confining unit
Selma chalk	Navarro Fm.	Confining unit
	Taylor Fm.	
Eutaw	Austin Fm.	Confining unit
	Eagle Ford Fm.	Saline reservoir
Tuscaloosa	Upper Tusc.	Minor reservoir
	Marine Tusc.	Confining unit
	Lower Tusc.	Saline reservoir (Injection zone)
Washita-Fredricksburg	Dantzler Fm.	Saline reservoir
	Limestone unit	

FIGURE 5.6
General Mississippi stratigraphy. (From Riestenberg, D. E. et al. 2009. *CO₂ Sequestration Permitting at the SECARB Mississippi Test Site*. San Antonio, TX: Society of Petroleum Engineers.)

transport facilities to inject CO_2 into a deep saline formation. The target storage formation is down-dip of the production horizon, located about 1,850 m below gas reservoir level, with thickness, porosity, and permeability of 20 m, 13%–20%, and 9.87e-15 m^2 (10 mD), respectively. The target formation is separated by 950 m of carboniferous mudstones (seal) from the production interval. During this project, it was planned that 17 million tons of CO_2 be injected into the underground storage via three injection wells (35).

5.5 CO$_2$ Storage in Citronelle Saline Aquifer: Simulation and Modeling

Injection and storage of CO_2 in Citronelle, AL, is phase III of the South Eastern Carbon Sequestration Partnership and aims to demonstrate commercial-scale storage of CO_2 captured from an existing coal-fired power plant. Alabama Power Company's Plant Barry is the source of CO_2 and is located approximately 19.3 km from the Southeast Citronelle Unit (Figure 5.7). The project will be capable of capturing approximately 125,000 metric tons of anthropogenic CO_2 per year. A pipeline was constructed from Plant Barry to Denbury's Southeast Citronelle Unit, and the CO_2 is injected into saline Paluxy sandstones at depths of approximately 3,048 m. Injection will continue for three years at a rate of 185,000 tons per year (t/year). After finishing injection, the sequestered CO_2 will be monitored for an additional four years to determine how well the CO_2 has been contained (42).

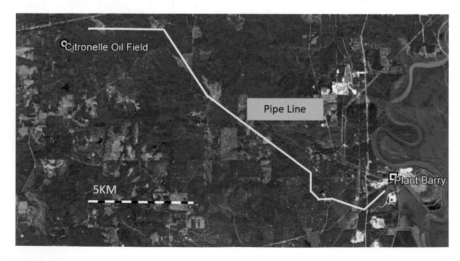

FIGURE 5.7
CCS project in the Citronelle field. (From Koperna, G. J. et al. 2012. The SECARB Anthropogenic Test: The First US Integrated Capture, Transportation, and Storage Test. *Carbon Management Technology Conference*, Orlando, FL.)

5.5.1 Geology of the Storage Formation

The CO_2 injection site is located inside the Citronelle oilfield boundaries on the southeast flank of Citronelle Dome, three miles from Mobile County in Alabama. The Citronelle oil field (discovered in 1955) covers more than 66.4 km² and includes about 500 wells. Water flooding, which started in 1961, has resulted in 270 MM m³ of cumulative oil production. The current average oil production is about 266 m³/day (43).

The Paluxy Formation (proposed injection zone) is located at a depth of about 2,865–3,200 m, and represents a coarsening-upward succession of variegated shale and sandstone (42). Based on logs from the injection well, 27 individual sandstones in the Paluxy Formation were identified as potential storage reservoirs for CO_2. The 17 thickest and most extensive sand layers were selected for injection. Citronelle Dome, a broad, gently dipping anticline, provides Citronelle field with structural closure at all stratigraphic horizons of Jurassic through Tertiary age, including Paluxy Formation (43). Moreover, there is an apparent lack of faulting at the Citronelle Dome structure. The proposed confining zone for this CO_2 injection test is the basal shale of the Washita-Fredericksburg interval and has an average thickness of 46 m (Figure 5.8). The aquifers on top and bottom of this confining unit (including Paluxy) exhibit extremely low groundwater velocities. Sand layers in the Paluxy Formation have satisfactory reservoir properties (porosity, permeability, and extent) for CO_2 storage.

Paluxy Formation, in the injection zone, can be divided into three sub-layers based on the thickness of the sand and shale layers. The middle section is mainly dominated by carbonate or limestone, while the others consist of thicker sand layers.

Cretaceous	Upper	Selma group			Confining unit
		Eutaw formation			Minor saline reservoir
		Tuscaloosa group	Upper		Minor saline reservoir
			Mid	Marine shale	Confining unit
			Lower	Pilot sand	Saline reservoir
Cretaceous	Lower	Washita-Fredericksburg interval	Dantzler sand		Saline reservoir
			Basal shale		Primary confining unit
		Paluxy formation	Upper		Injection zone
			Middle		
			Lower		
		Mooringsport formation			Confining unit
		Ferry lake anhydrite			Confining unit

FIGURE 5.8
Stratigraphic column for Citronelle field. (From Denbury Resources, Incorporated Plan. 2010. *SECARB Phase III Anthropogenic Test Volume 1 of 2*. Montgomery, AL: Alabama Department of Environmental Management.)

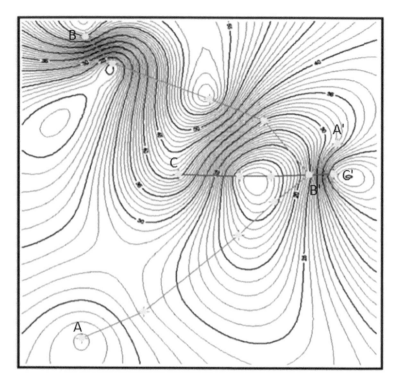

FIGURE 5.9
Location of three cross sections in the Citronelle Dome.

The geological model of the Paluxy Formation is based on the interpretation of 16 well logs in three cross sections (Figure 5.9). Injection well D-9-7 is the well considered as a reference well in three cross sections. Seventeen sand layers are picked and correlated (Figure 5.10) based on the highest resistivity and highest spontaneous potential (SP) values. The areal dimensions of some of the thicker sandstones are on the order of 15.5 square miles or 1,554 hectares. The total thickness of the sand layers is about 143 m, ranging from 3 to 24 m. Structural and isopach maps of the sand layers are used to make the reservoir simulation model.

The final structure maps for the top of the each formation and thickness of 17 sand layers after being imported into the reservoir simulator are depicted in Figures 5.11 and 5.12.

5.5.2 Reservoir Simulation Model

Based upon the interpretation and evaluation of geophysical well logs, a comprehensive picture of the subsurface geology was developed for the reservoir simulation modeling. The reservoir models were built using Computer Modeling Group (CMG) software. The geological structure of the

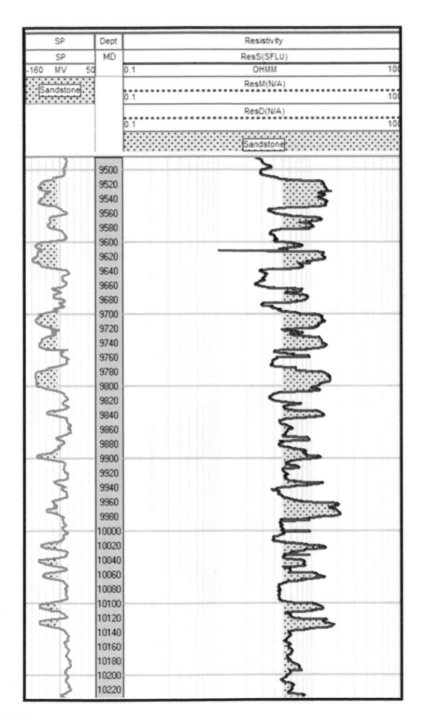

FIGURE 5.10
Sand layers in the injection well or D-9-7 (SI conversion: 1 ft = 0.3 m).

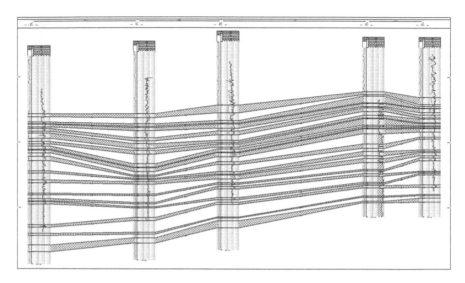

FIGURE 5.11
Correlated sand layers of cross section B–B'.

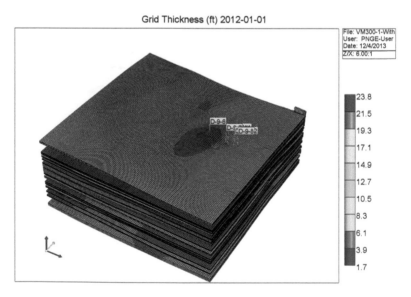

FIGURE 5.12
Thickness maps for sand layers (SI conversion: 1 ft = 0.3 m).

model includes 17 sand layers representing 51 simulation layers. The Cartesian grid coordinates had dimensions of 150 × 150 × 51 grids (Δx and Δy equal to 40.5 m). Local grid refinement was considered around the injection well. Well logs from 40 offset wells within the area of study were acquired and interpreted to generate porosity maps. Resistivity logs were used for the injection depth, which is in the range of about 2,865 m to 3,200 m. The Archie equation was used (43) to calculate the porosity values using the resistivity logs:

$$\phi = \left(\frac{a}{(R_t/R_w)S_w{}^n} \right)^{1/m}$$

where

a = tortousity factor = 1 (default value)
m = cementation factor = 2.25 (best match to Citronelle oilfield porosity logs)
n = saturation exponent = 2 (common default value)
R_w = resistivity of the formation water = 0.045 (best match to Citronelle oilfield porosity logs)
S_w = water saturation = 0.95 (assuming only residual gas saturation)
ϕ = porosity
R_t = true formation resistivity (obtained from logs).

The weighted average of the porosity values was calculated by taking the thickness h of each layer into account: $\phi_{ave} = \sum \phi h / \sum h$.

The porosity map for the first simulation layer is shown in Figure 5.13. Porosity–permeability cross plots obtained from core analysis provide reasonable estimates for the permeability distribution within the reservoir.

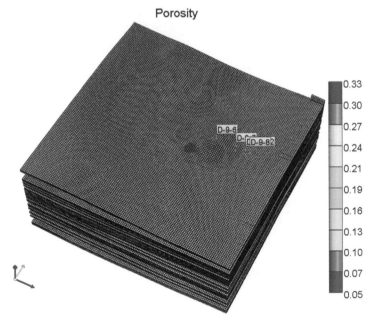

FIGURE 5.13
Porosity map for the first layer generated by 40 well logs.

Relative permeability curves (Figure 5.14) from history matching the injection pilot test at the Mississippi Test Site were used in this simulation (43). The trapped gas saturation was considered to be 7.5%; this value determines how much gas is trapped residually due to hysteresis effects. For the reservoir simulation, the injection well was operated with a maximum bottom-hole pressure limit of 43.4×10^3 kPa (based on a conservative fracture pressure gradient of 1.62 kPa/m (41)) and an injection rate constraint of 0.27 million standard cubic meters per day (500 t/day). Injection started at the beginning of 2012 and took three years. The saline formation was assumed to be a closed reservoir (no-flow boundary condition). Other reservoir properties are summarized in Table 5.2. This is considered as a base case model in the following sections.

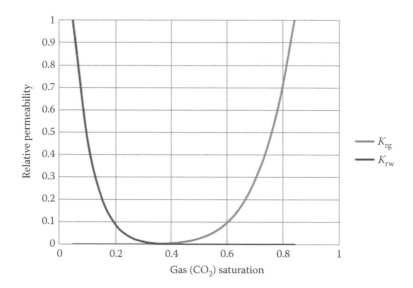

FIGURE 5.14
Relative permeability curves used in the reservoir simulation model.

TABLE 5.2

Reservoir Parameters and Properties (Base Case Model)

Parameter	Value	Parameter	Value
Permeability (mD)	0.824 exp 28.18ϕ	Water density (kg/m³)	993
Temperature (°C)	110	Water viscosity (cP)	0.26
Salinity (ppm)	100,000	Water compressibility (1/bar)	4.64E-05
Residual gas saturation	0.35	K_v/K_h (permeability ratio)	0.1
Residual water saturation	0.6	Pressure reference @2,870 m (kPa)	3.03E+04

Note: ϕ is porosity.

The initial reservoir simulation runs showed that the maximum extent of the CO_2 plume takes place in the first (top) layer and sixth layer. This is mainly due to the fact that these layers represent sands with higher permeability, which causes CO_2 to migrate further from the injection well. As shown in Figure 5.15, the plume area is expected to have an approximate major diameter of 1,504 m, 500 years after the end of injection (42). The CO_2 plume extent in all target layers (vertically) is shown in Figure 5.16.

Two pressure down-hole gauges (PDGs) were installed in the project's observation well (D-9-8#2), which is located 250 m from the east side of the injection well. These PDGs can provide real-time pressure and temperature measurements. The actual pressure data can be used for reservoir monitoring (especially CO_2 leakage detection) in addition to history matching. Therefore, the main focus of this study is to analyze the reservoir simulation pressure behavior at the grid block that corresponds to the exact location of the PDG in the observation well. Pressure in the observation well rose from 30.33 to 32.6 MPa (maximum pressure) during the three years of injection from 2012 to 2015. After the injection was stopped, the pressure decreased gradually to 32.13 MPa after one year (stabilized pressure). Finally, the reservoir pressure in the observation well followed a very gentle decline and stable trend from 32.13 to 32.08 MPa over 500 years (Figure 5.17).

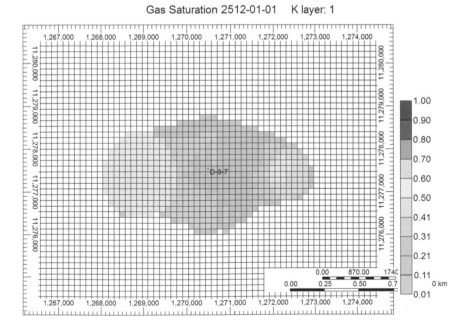

FIGURE 5.15
CO_2 plume extension in the first layer 500 years after injection (SI conversion: 1 ft = 0.3 m).

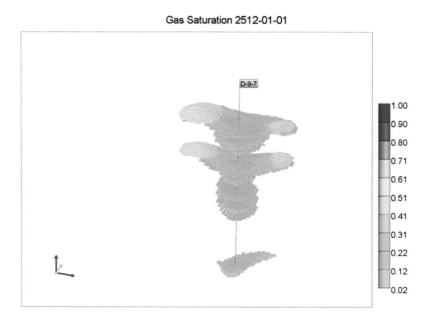

FIGURE 5.16
CO$_2$ plume extension in all layers 500 years after injection.

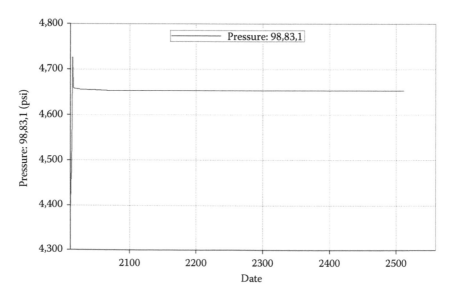

FIGURE 5.17
Pressure behavior in the observation well, base case mode l (SI conversion: 1 psi = 6.9 kPa).

5.6 Reservoir Characteristics Sensitivity

In this section, simulation model predictions are presented while considering uncertainty in some reservoir properties. The sensitivity analysis procedure changed one parameter at a time within the uncertainty range to investigate the corresponding effects on reservoir pressure at the observation well and the CO_2 plume extent (45,46). The reservoir parameters analyzed in this section are permeability (rock type), gas relative permeability, maximum residual gas saturation (hysteresis), and vertical to horizontal permeability ratio, boundary condition, brine compressibility, and density. Since the CO_2 plume extent shape is elliptical, the magnitude of the major and minor axes (Figure 5.15) can characterize the underground CO_2 distribution 5, 50, or 500 years after injection. Additionally, to analyze the reservoir pressure behavior, we focused on the maximum (at the end of injection) and stabilized pressures.

5.6.1 Permeability

From here on, the contribution of each parameter to the reservoir pressure and plume extent are identified. In the Citronelle reservoir model, the porosity originates from maps generated by the interpretation of 40 well logs. Figure 5.18 shows porosity–permeability cross plots of the Geological Survey of Alabama's Southwestern Dataset (43). In order to have a reliable

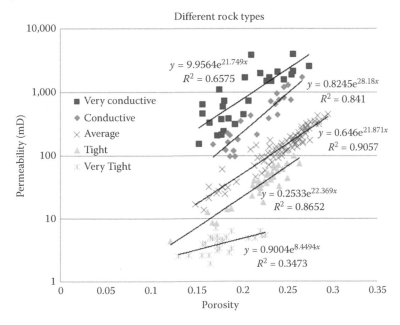

FIGURE 5.18
Porosity–permeability cross plot (SI conversion: 1 D = 9.87e-13 m²).

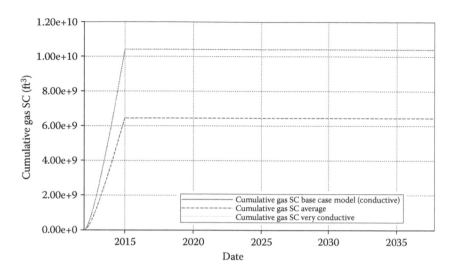

FIGURE 5.19
CO$_2$ injectivity for different rock types (SI conversion: 1 ft^3 = 0.028 m^3).

porosity–permeability correlation, the data points are clustered into five different rock types, ranging from very tight to very conductive. The initial porosity–permeability data gathered from well D-9-8 (observation well) core analysis represents a Conductive rock type (this is used in the base model). Average ($K = 0.64e^{21.87\phi}$) and Very Conductive ($K = 9.964e^{21.74\phi}$) rock types are introduced into the reservoir simulation model.

For the Average rock type, due to the lower permeability values, CO$_2$ injectivity decreases, and it is therefore not possible to store all the CO$_2$ according to the planned target (Figure 5.19). The injectivity of CO$_2$ is the same for both Conductive and Very Conductive rock types and is equal to the target values. We can see the results for pressure in Figure 5.20. The stabilized reservoir pressure changes very gently over 500 years. Here, results for the 20 years after injection are shown to provide more detail. By decreasing the permeability (use of Average rock type), CO$_2$ injectivity decreases to 60% of the target value, resulting in a reduced reservoir pressure compared with the base case. With higher permeability (Very Conductive rock type), the stabilized reservoir pressure is 290 kPa less than the base case due to the higher conductivity, which prevents more pressure build up. Additionally, an increase in the permeability enhances CO$_2$ and brine displacement, which leads to larger CO$_2$ plume extents, according to Table 5.3.

5.6.2 Permeability Ratio

Typically, vertical permeability is determined as a ratio with horizontal permeability. In this study, for the base case model, K_v/K_h is considered to be 0.1. For the sensitivity analysis we assigned values of 0.3, 0.5, and 0.7 to K_v/K_h. As shown in Figure 5.21, an increase in K_v/K_h generates less pressure build

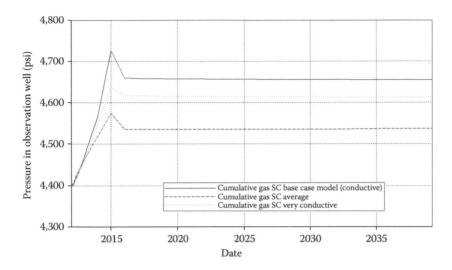

FIGURE 5.20
Reservoir pressure in observation well for different rock types (SI conversion: 1 psi = 6.9 kPa).

TABLE 5.3

CO$_2$ Plume Extension Size over Time (in the First Layer) for Different Rock Types

			Permeability		
			Base Case	Average	Very Conductive
			0.824 exp 28.18ϕ	0.640 exp 21.87ϕ	9.96 exp 21.74ϕ
CO$_2$ plume extension	5 years after injection	Minor axis (m)	650	488	772
		Major axis (m)	732	528	975
	50 years after injection	Minor axis (m)	772	569	853
		Major axis (m)	1,219	650	1,423
	500 years after injection	Minor axis (m)	813	650	813
		Major axis (m)	1,504	1,138	1,544

Note: ϕ is porosity.

up during injection. However, after the transition time, the higher the vertical to horizontal permeability ratio, the higher the stabilized pressure value is. Also, the size of the CO$_2$ plume increases slightly for higher K_v/K_h, especially for 5 and 50 years after injection (Table 5.4).

5.6.3 Gas Relative Permeability Curves

Four different gas relative permeability curves were generated, with two representing higher values of relative permeability at any given gas saturation and two representing lower values compared with the base case (Figure 5.22).

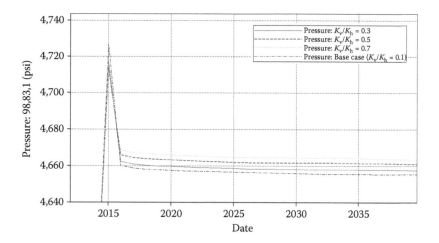

FIGURE 5.21
Reservoir pressure in observation well for different permeability ratios (SI conversion: 1 psi = 6.9 kPa).

TABLE 5.4

CO_2 Plume Extension Size over Time (in the First Layer) for Different Permeability Ratios

			Vertical to Horizontal Permeability Ratio			
			Base Case	$K_v/K_h = 0.3$	$K_v/K_h = 0.5$	$K_v/K_h = 0.7$
CO_2 plume extension	5 years after injection	Minor axis (m)	650	650	650	691
		Major axis (m)	732	772	813	853
	50 years after injection	Minor axis (m)	772	813	813	813
		Major axis (m)	1,219	1,260	1,301	1,341
	500 years after injection	Minor axis (m)	813	813	853	853
		Major axis (m)	1,504	1,504	1,544	1,544

It is worth mentioning that the curves with higher gas relative permeability values have lower residual gas saturations and vice versa. Results are shown in Table 5.5 and Figure 5.23. The higher gas relative permeability curves correspond to lower residual gas saturations, which can mobilize the CO_2 phase earlier (at lower gas saturations). Therefore, CO_2 moves further, resulting in a larger CO_2 plume extent. Additionally, a higher gas relative permeability increases the stabilized reservoir pressure. Conversely, a lower gas relative permeability leads to a less extensive plume and lower stabilized reservoir pressure.

5.6.4 Maximum Residual Gas Saturation

Generally, drainage relative permeability curves are provided for the reservoir simulation model. When the maximum residual gas saturation is

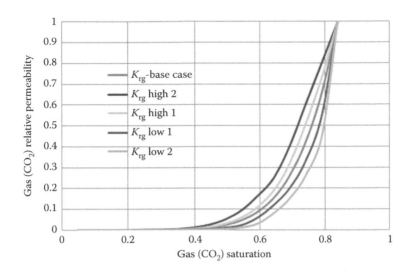

FIGURE 5.22
Different gas relative permeability curves.

TABLE 5.5

Plume Extension Size over Time in the First Layer for Different Gas Relative Permeability Curves

			Gas Relative Permeability K_{rg}				
			Base Case	Low 2	Low 1	High 1	High2
CO$_2$ plume extension	5 years after injection	Minor axis (m)	650	569	610	650	732
		Major axis (m)	732	650	691	772	853
	50 years after injection	Minor axis (m)	772	650	691	813	894
		Major axis (m)	1,219	853	1,057	1,301	1,382
	500 years after injection	Minor axis (m)	813	732	772	853	894
		Major axis (m)	1,504	1,260	1,382	1,544	1,666

introduced, the imbibition gas relative permeability curve can be determined based on the drainage curve (51). During CO$_2$ movement in the reservoir, water imbibition causes a portion of the gas phase to be trapped in the pores (residual trapping). Therefore, when the maximum residual gas saturation increases, more gas is trapped, resulting in less mobile CO$_2$ and consequently a smaller CO$_2$ plume extension (Table 5.6). Changing the maximum residual gas saturation has no significant impact on reservoir pressure.

5.6.5 Brine Compressibility

In a closed geological system, the amount of CO$_2$ that can be injected into the saline reservoir is mostly dependent on the availability of the additional pore

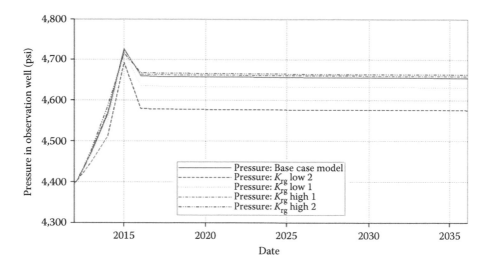

FIGURE 5.23
Reservoir pressure in the observation well for different gas relative permeability curves
(SI conversion: 1 psi = 6.9 kPa).

TABLE 5.6

CO_2 Plume Extension Size over Time in the First Layer for Different Maximum
Residual Gas Saturations

			Maximum Residual Gas Saturation (Hysteresis)			
			Base Case	**0.05**	**0.1**	**0.2**
CO_2 plume extension	5 years after injection	Minor axis (m)	650	650	650	650
		Major axis (m)	732	732	732	732
	50 years after injection	Minor axis (m)	772	772	772	732
		Major axis (m)	1,219	1,219	1,179	1,138
	500 years after injection	Minor axis (m)	813	853	772	772
		Major axis (m)	1,504	1,504	1,463	1,382

space that can be provided due to brine compressibility (46). Additionally,
the compressibility determines how much injected fluid contributes to
reservoir pressure build up or brine volume change (this can also be referred
to as a change in brine density). As observed in Figure 5.24, an increase in
brine compressibility results in a lowering of the maximum and stabilized
reservoir pressures. With higher brine compressibility, injected CO_2 results in
more changes in brine density rather than generating pressure build up in the
reservoir. Changing the brine compressibility has no considerable influence
on the CO_2 plume extent.

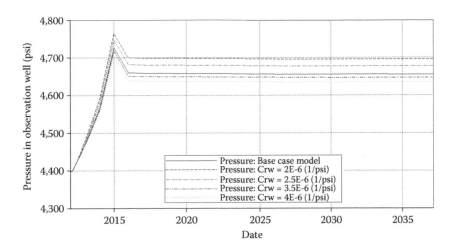

FIGURE 5.24
Reservoir pressure in the observation well for different brine compressibilities (SI conversion:
1 psi = 6.9 kPa).

5.6.6 Brine Density

The impact of a change in brine density on reservoir pressure can be analyzed
by considering the fact that the denser the brine, the less compressible it
is, allowing more pressure build up during and after CO$_2$ injection. As is
illustrated in Figure 5.25, a higher brine density contributes more pressure
gain for the reservoir (both maximum and stabilized pressures). The influence

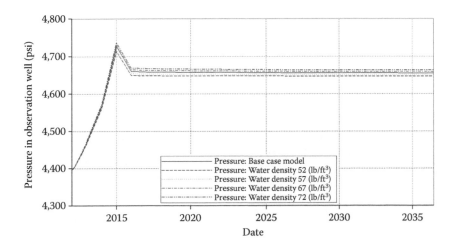

FIGURE 5.25
Reservoir pressure in the observation well for different brine densities (SI conversion:
1 psi = 6.9 kPa).

TABLE 5.7

CO₂ Plume Extent Size over Time (in the First Layer) for Different Brine Densities

			Brine Density (lb/ft³)				
			Base Case	833	913	1,073	1,153
CO₂ plume extension	5 years after injection	Minor axis (m)	650	650	650	650	650
		Major axis (m)	732	691	732	772	813
	50 years after injection	Minor axis (m)	772	772	772	772	813
		Major axis (m)	1,219	1,057	1,138	1,260	1,260
	500 years after injection	Minor axis (m)	813	772	813	853	853
		Major axis (m)	1,504	1,423	1,504	1,504	1,544

of brine density on CO₂ plume extent is addressed by the driving mechanism that governs fluid movement in the reservoir. During CO₂ injection, viscous forces make the CO₂ move forward, and, after injection, buoyancy is the dominant driving force. The density difference between the brine and CO₂ determines the magnitude of the buoyant force (52). A higher brine density results in a greater density differential and, consequently, more buoyancy force. Therefore an increase in brine density means more buoyancy force is exerted to the CO₂ plume, resulting in larger extents (Table 5.7).

5.6.7 Boundary Condition

In this section, we assume that the saline reservoir in the Paluxy Formation is not a closed system. A Fetkovich aquifer, where the reservoir pressure is kept constant at the reservoir boundaries, is assigned to the east, east-south and east-south-west edges of the reservoir (Figure 5.26). As shown in Figure 5.27, the reservoir pressure behavior in the open system is significantly different than observed in the previous sections. First of all, the maximum reservoir pressure at the end of injection is much less (almost 1,380 kPa) in the open system. Secondly, the stabilized pressure reverts to the initial or native reservoir pressure at some point.

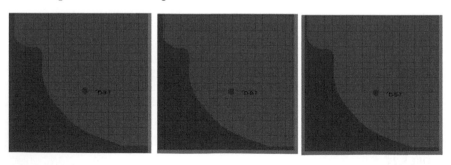

FIGURE 5.26
Different locations for constant pressure boundary (Fetkovich aquifer).

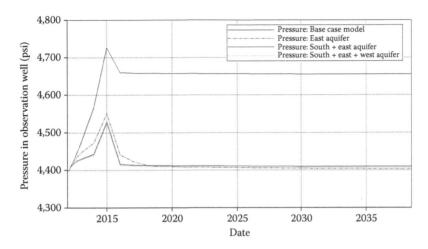

FIGURE 5.27
Reservoir pressure in the observation well for different boundary conditions (SI conversion: 1 psi = 6.9 kPa).

As the portion of the reservoir boundary that is exposed to constant pressure (Fetkovich aquifer) increases, less pressure build up is observed at the end of injection. Additionally, when more edges of the reservoir are connected to the open aquifer, it takes less time for the reservoir pressure to reach native conditions. Changing the reservoir boundary conditions has an insignificant effect on CO_2 plume size (Table 5.8).

The results of sensitivity analysis can be considered for risk assessment in addition to history matching the reservoir simulation model if actual field measurements (pressure data) are available. The main findings can be summarized as follows:

- The rock type (permeability) contributes significantly to CO_2 injectivity, reservoir pressure, and CO_2 plume extent. Higher permeability leads to a more extensive CO_2 plume and less reservoir

TABLE 5.8

Different Locations for Constant Pressure Boundary (Fetkovich Aquifer)

			Reservoir Boundary (Fetkovich Aquifer)			
			Base Case	East	East + South	East + South + West
CO_2 plume extension	5 years after injection	Minor axis (m)	650	650	691	691
		Major axis (m)	732	732	813	853
	50 years after injection	Minor axis (m)	772	772	813	813
		Major axis (m)	1,219	1,219	1,260	1,260
	500 years after injection	Minor axis (m)	813	853	853	853
		Major axis(m)	1,504	1,544	1,544	1,544

pressure gain. An increase in vertical to horizontal ratio leads to higher stabilized pressure and CO_2 plume extent.

- An increase in gas relative permeability results in a higher stabilized pressure and a larger CO_2 plume extent. Additionally, higher maximum residual gas saturation leads to more residual trapping, and a lower CO_2 plume extent.

- Brine compressibility plays a role in reservoir pressure build up, especially in a closed geological system. When brine compressibility rises, we observe a decrease in stabilized reservoir pressure.

- Density of brine affects both reservoir pressure and CO_2 plume extent. Denser brine causes more buoyancy force, which drives CO_2 to move further and distribute in a greater area. Higher brine density values also contribute to more reservoir pressure build up.

- Changing the boundary condition of the reservoir from closed to constant pressure affects reservoir pressure behavior significantly. When Fetkovich aquifers are placed at the edges of the reservoir, the maximum pressure build up decreases notably. In addition, when injection ceases, the stabilized reservoir pressure reverts to the native reservoir pressure after a while.

5.7 Trapping Mechanisms

The simulation and modeling techniques used to describe the behavior of the four main trapping mechanisms (structural, residual, solubility, and mineral trapping) are those described by L. Nghiem (47) when studying the simulation and optimization of trapping processes for CO_2 storage in saline aquifers. Four scenarios based on different relative permeability profiles were defined in order to study its effects on the contribution that each of the four main trapping mechanisms provide during the CO_2 sequestration process. This will finally allow determination of the "safest" and "riskiest" scenario in terms of CO_2 mobility.

The relative permeability profiles chosen for this study correspond to experimental data for supercritical CO_2/brine systems from the West Canadian Sedimentary Basin describing both drainage and imbibition processes (48). These parameters are the values that the GEM simulator will use for its rock–fluid data interpretation (Figures 5.28 and 5.29).

5.7.1 Structural Trapping

The single most important factor for securing CO_2 is the presence of a thick and fine-textured rock to serve as a seal above the sequestration reservoir.

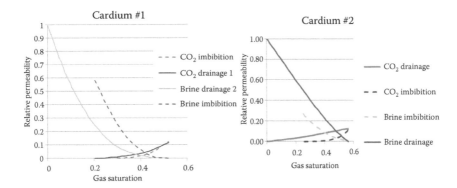

FIGURE 5.28

Drainage and imbibition relative permeability profiles experimental data for supercritical CO_2/brine systems from the West Canadian Sedimentary Basin.

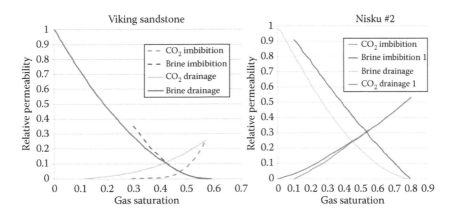

FIGURE 5.29

Drainage and imbibition relative permeability profiles experimental data for supercritical CO_2/brine systems from the West Canadian Sedimentary Basin.

The seal should provide an effective permeability and capillary barrier to upward migration. The effects of the properties of this seal or confining unit are studied thoroughly in Section 5.8.

Once injected, the supercritical CO_2 may be more buoyant than other liquids present in the pore space. The CO_2 will therefore percolate up through the porous rocks until it reaches the top of the formation where it meets (and is trapped by) an impermeable layer (confining unit).

5.7.2 Residual or Capillary Gas Trapping Modeling

Simulating and modeling residual trapping was performed using Land's residual gas trapping model (49) (Figure 5.30). When CO_2 saturation increases,

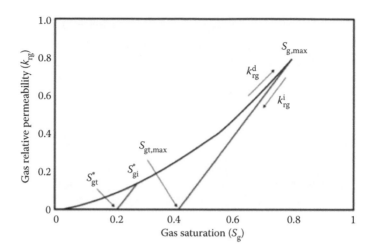

FIGURE 5.30
Land's residual gas trapping model.

the gas relative permeability follows the drainage curve (black). If at any given saturation (S_{gi}*) on the drainage curve, the gas saturation reverses its course and decreases, the gas follows the imbibition curve (red), which shows the irreversibility of the process (hysteresis).

Land's coefficient C is defined as

$$C = \frac{1}{S_{gt,max}} - \frac{1}{S_{g,max}}$$

where $S_{g,max}$ refers to the maximum gas saturation that can be attained, and $S_{gt,max}$ is the maximum trapped gas saturation. The residual gas saturation for any given gas saturation (S_{gi}*) is defined as

$$S_{gt}^*\left(S_{gi}^*\right) = \frac{S_{gi}^*}{1 + CS_{gi}^*}$$

5.7.3 Solubility Trapping Modeling

The second CO$_2$ storage mechanism is CO$_2$ dissolution in reservoir water. CO$_2$ is soluble in water, and when injected into a pressurized saline reservoir, some of the CO$_2$ will dissolve in the formation water. The amount of CO$_2$ ultimately dissolved in water is affected by several factors, including temperature and pressure within the reservoir, salinity of the reservoir water, and reservoir heterogeneity and geometry.

CO$_2$ solubility in brine is modeled as a phase-equilibrium process, which is governed by the equality of fugacities in the gas and aqueous phase:

$$f_{ig} = f_{ia}, \quad i = 1, \ldots n_c$$

The fugacity of component "i" in the gas phase (f_{ig}) is calculated with the equation of state (Peng-Robinson). In this model the fugacity of component "i" in the aqueous phase (f_{ia}) is modeled with Henry's laws:

$$f_{ia} = H_i \times x_{ia}$$

Henry's constant (H_i) is a function of pressure, temperature, and water salinity, which has to be input in addition to basic water properties (density, compressibility, and viscosity). A survey of Henry's constant correlations for important gases associated with greenhouse gas (GHG) sequestration in saline reservoirs was conducted. In 1996, Harvey (50) published correlations to determine Henry's constants for many gaseous components including CO$_2$, N$_2$, H$_2$S, and CH$_4$. These correlations have been implemented in GEM-GHG and the default values were used in the simulation.

5.7.4 Mineral Trapping Modeling

The final CO$_2$ storage method involves mineral trapping. Mineral trapping is the permanent sequestration of CO$_2$ through chemical reactions with dissolved minerals in the reservoir brine and with the minerals in the reservoir rock itself. However, the mineral trapping mechanism is slow and is expected to occur over very long time periods. Through field studies and numerical modeling, it has been determined that CO$_2$ is primarily trapped through precipitation of calcite (CaCO$_3$), siderite (FeCO$_3$), dolomite (CaMg(CO$_3$)), and dawsonite (NaAlCO$_3$(OH)$_2$). In order for mineral trapping through carbonate precipitation to occur, primary minerals rich in Mg, Fe, Na, and Ca, such as feldspars and clays, must be present. Therefore, immature sandstones with an abundance of fresh rock fragments (unweathered igneous and metamorphic minerals and clays rich in Mg, Fe, and Ca) are most effective (Bachu et al. (1994); Preuss et al. (2001); Xu et al. (2002)). The abundance and ratios of these primary minerals can have a tremendous effect on the type of secondary minerals that are precipitated as well as on the overall total amount of CO$_2$ sequestered through mineral trapping.

The GEM simulator, with the use of the greenhouse gas module (GHG), can evaluate the chemical reactions among chemical species in the aqueous phase and the dissolution/precipitation of minerals. To describe the chemical reactions, it is useful to define the number of chemical reactions and the number of components in the gas, aqueous, and mineral phases. This process is thoroughly described by Nghiem et al. (47).

Despite the fact that when creating this model no mineralogy data was available, the following (most common) reactions expected in sandstone were modeled and are summarized as follows:

$$CaSO_4(\text{anhydrite}) = Ca^{2+} + SO_4^{2-}$$

$$CaCO_3(\text{calcite}) + H^+ = Ca^{2+} + HCO_3^-$$

$$CaMg(CO_3)_2(\text{dolomite}) + 2H^+ = Ca^{2+} + Mg^{2+} + 2HCO_3^-$$

$$FeCO_3(\text{siderite}) + H^+ = Fe^{2+} + HCO_3^-$$

$$NaAl(CO_3)(OH)(\text{dawsonite}) + 3H^+ = Na^+ + Al^{3+}HCO_3^- + 2H_2O$$

5.7.5 Trapping Mechanisms Sensitivity

After performing the simulations for a period of 500 years, a sensitivity analysis was performed on the effects of different relative permeability profiles on the contribution and behavior of each one of the four main trapping mechanisms (residual, structural, solubility, and mineral trapping). The four different scenarios are summarized in Table 5.9.

After thorough analysis of the presented results we can observe that, even though the profiles differ in ranges of relative permeability, CO$_2$ critical saturations and CO$_2$ residual gas saturation, the main parameter that controls the process behavior is the residual gas saturation. If the residual saturations are higher, this will lead to higher amounts of residually or capillary trapped CO$_2$, as can be observed in Figure 5.31. This behavior can also be observed in terms of dissolved CO$_2$, as shown in Figure 5.32.

The contribution that each trapping mechanisms provides to the CO$_2$ storage process was quantified. This is possible by knowing the total amount of CO$_2$ injected into the formation (426 MM m^3 or 550,596.35 t CO$_2$) and

TABLE 5.9

Relative Permeability Profiles Summary

Scenario	Relative Permeability Profile	Residual Gas Saturation (%)	Critical Gas Saturation (%)	Maximum K_{rg}
1	Cardium #1	30.090	20.370	0.117
2	Cardium #2	28.980	1.080	0.122
3	Viking Sandstone	25.810	11.100	0.258
4	Nisku #2	10.660	2.930	0.533

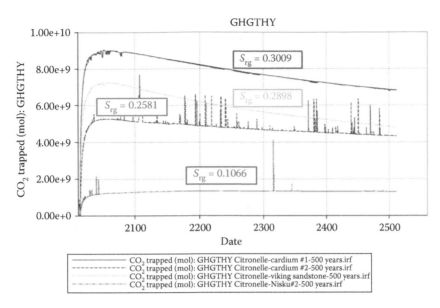

FIGURE 5.31
Comparison of amount of capillary trapped CO_2 for the different relative permeability profiles – 500 years.

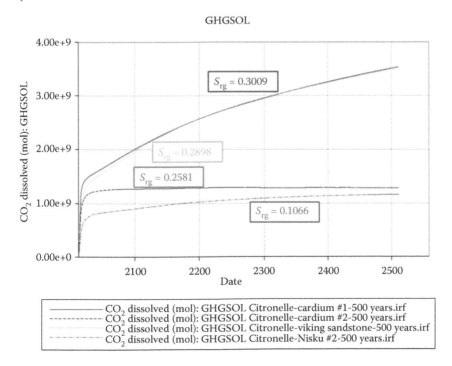

FIGURE 5.32
Comparison of amount of dissolved CO_2 for the different relative permeability profiles 500 years.

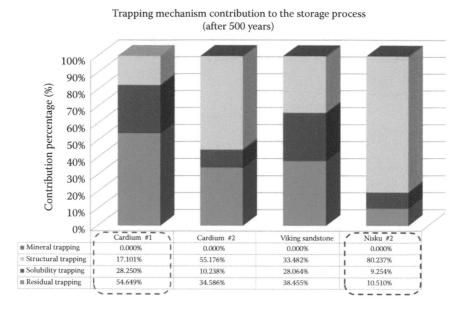

FIGURE 5.33

Trapping mechanisms contribution – 500 years.

comparing it to the amounts of CO_2 trapped by the different mechanisms (Figure 5.33). Mineral trapping is also included, but does not show a significant contribution to this process in the first 500 years, and detailed knowledge of the mineralogy of the formation is necessary to make a proper assessment of the contribution of this mechanism.

We can also observe which relative permeability profile will represent a "safer" scenario in terms of CO_2 storage in the case where a leakage occurs. The relative permeability profile corresponding to Scenario 1 (Cardium #1) results in 54.7% of the CO_2 being residually or capillary trapped, 28.25% dissolved in the brine, and 17.1% structurally trapped, implying that more than 80% of the CO_2 will remain immobile in the formation in the event of failure of the confining unit or any other means of leakage. In the relative permeability profile corresponding to Scenario 4 (Nisku #2), more than 80% of the CO_2 is structurally trapped and only 20% of the injected amount is trapped by different mechanisms, representing a scenario where 80% of the injected CO_2 will still be mobile and able to migrate in the event of a leakage.

5.8 Seal Quality Analysis

The main confining unit of the Paluxy Formation is within the Washita-Fredericksburg interval, as mentioned in Section 5.5.1, where it is thoroughly described. The Washita-Fredericksburg interval is formed by the Danztler Sand, which is an aquifer with characteristics similar to those of the Paluxy Formation, and the Basal Shale as the main confining unit. The quality and integrity of the cap-rock, which provides the ability to trap "mobile" CO_2 structurally, is studied to assess the potential risk of leakage, as well as the impact that the properties of this confining unit may have on the pressure and saturation distributions within the Paluxy Formation.

In order to perform this analysis, a new cross-section was generated including wells near the area of interest (D-9-3, D-9-7, and D-9-9), as shown in Figure 5.34, to assess the properties of the seal. The Washita-Fredericksburg interval is located at depths between 2,243 and 2,865 m. By using the SP log it was possible to identify the top and bottom of the Basal Shale, finding that throughout the area of interest the thickness of this shale ranged between 52 and 73 m (Figure 5.35).

Finally, these two additional geological layers (Basal Shale and Danztler Sand) were included in the base model corresponding to the formations within the Washita-Fredericksburg interval (Figure 5.36). Each geological layer was then initially divided into three simulation layers.

A sensitivity analysis was performed in terms of the thickness and permeability of the Basal Shale. Nine (9) different scenarios were designed, with thickness ranging between 150 and 250 ft and permeability within

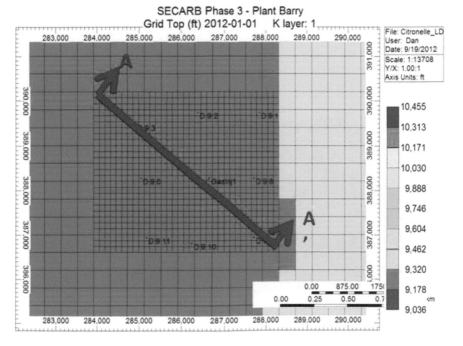

FIGURE 5.34
Cross-section A–A' (SI conversion: 1 ft = 0.3 m).

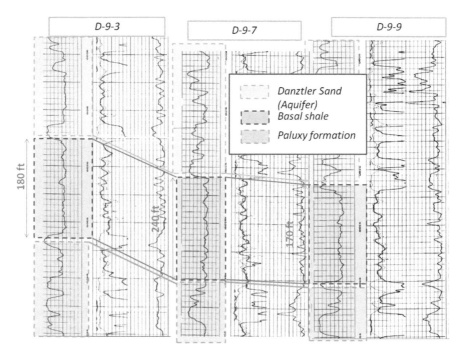

FIGURE 5.35
Well log interpretation of cross-section A–A′ (SI conversion: 1 ft = 0.3 m).

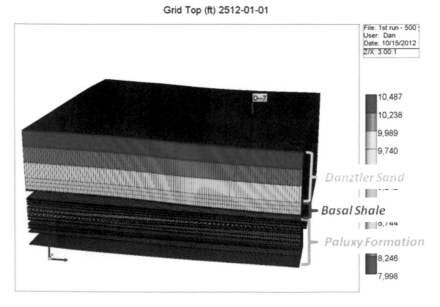

FIGURE 5.36
Inclusion of the Washita-Fredericksburg geological layers (SI conversion: 1 ft = 0.3 m).

TABLE 5.10

Scenarios for Sensitivity Analysis on the Effects
of the Basal Shale Permeability and Thickness

Scenarios	Thickness (m)	Permeability (m²)
1	46	1.00E-14
2	46	1.00E-16
3	46	1.00E-18
4	61	1.00E-14
5	61	1.00E-16
6	61	1.00E-18
7	76	1.00E-14
8	76	1.00E-16
9	76	1.00E-18

1E-14 m² (10^{-3} darcy) and 1E-14 m² (10^{-7} darcy) for a period of 500 years. These scenarios are summarized in Table 5.10.

Based on the initial results of this simulation, it was determined that the vertical resolution needed to be increased when modeling the confining unit. Two particular factors made this necessity quite evident. The first factor was the maximum amount of CO_2 present within the seal layers. Since this geological layer was divided into three simulation layers, the dimensions of the grid blocks in the vertical direction were between 15 and 25 m depending on the scenario. This did not allow the vertical gradual changes in the CO_2 mole fraction to be perceived within the confining unit; hence, a proper assessment of the depth of invasion of the CO_2 within this confining unit could not be performed. The second factor corresponds to the migration behavior of the CO_2 within the confining unit, which suggested that this migration would initially only occur vertically from the well grid block, and then move horizontally within the shale. After thorough analysis, it was decided to locally refine the grid blocks that corresponded to the confining unit in order to better grasp the complexity of the process taking place. This refinement was made such that the vertical dimensions of the grid blocks would not exceed 0.6 m, and simulation of the nine previously defined scenarios was performed (Figure 5.37).

When performing the sensitivity analysis based on the properties of the confining unit (thickness and permeability) two different aspects were studied. The first corresponds to the depth of invasion, which is the maximum depth (within the seal) where CO_2 is present, and acts as an indicator of how the confining unit has been compromised by "allowing" the migration of injected CO_2 within itself. As mentioned earlier, nine different scenarios were defined in order to perform this analysis, and the CO_2 mole fraction was used to assess CO_2 invasion within the confining unit. According to the results, as permeability decreases, so will the invasion of CO_2 within the confining unit. It is appropriate to mention that the high permeability scenario (1×10^{-14} m²) is not characteristic of shale formations, particularly a formation that would be selected as a proper confining unit when selecting CO_2 storage formation targets.

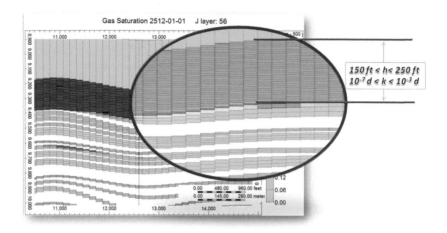

FIGURE 5.37
Grid refinement of the simulation layers of the confining unit (SI conversion: 1 ft = 0.3 m, 1 D = 9.87e-13 m^2).

After analyzing the results for all the scenarios available, it was observed that permeability is the main factor that determines the depth of invasion within the confining unit, with the exception of those scenarios considered to have high permeability (1×10^{-14} m^2). In these scenarios, the confining unit is fully breached by the injected CO$_2$ and the depth of invasion will then be the same as the thickness of the seal. These results are summarized in Figure 5.38; as mentioned earlier, in high permeability scenarios the depth of invasion ranges between 46 and 76 m (full seal breach), in contrast with scenarios where the permeability is 1×10^{-16} and 1×10^{-18} m^2, where the maximum depth of invasion is 6.7 and 2.4 m, respectively.

The second aspect analyzed in the context of this study was the impact that variations in the properties of the confining unit have on saturation and pressure distribution within the first layer of the Paluxy Formation. This layer is the most extensive and conductive layer in the Paluxy Formation, and is also the layer in permanent contact with the confining unit.

Regarding the saturation distribution in the first layer of the Paluxy Formation, it was observed that a change in the permeability of the confining unit has the most significant impact, as opposed to variations in thickness. This is due to the fact that, when in the presence of a quite conductive confining unit, changes in the saturation profiles are mainly driven by "fluid loss", shown by the ability of the CO$_2$ to migrate within the confining unit; as the permeability of the confining unit decreases, less CO$_2$ will be able to migrate upwards within the seal, so a higher concentration of CO$_2$ will remain in the Paluxy Formation, mainly migrating horizontally once it reaches this "impermeable" layer. This behavior was observed in the nine scenarios defined for this study (Figure 5.39).

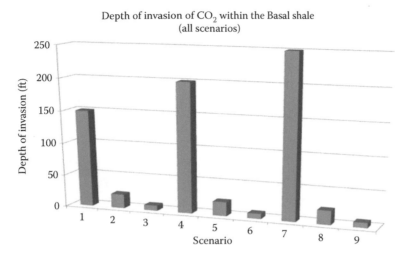

FIGURE 5.38
Depth of invasion within the confining unit – all scenarios (Table 5.10) comparison (SI conversion: 1 ft = 0.3 m).

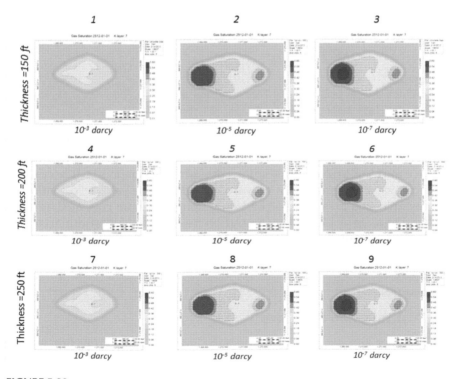

FIGURE 5.39
Saturation distribution in the Paluxy Formation top layer after 500 years – all scenarios (SI conversion: 1 ft = 0.3 m, 1 D = 9.87e-13 m²).

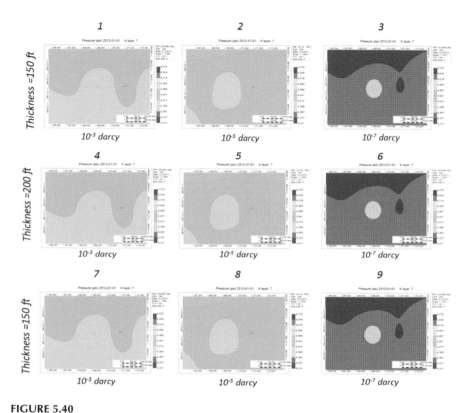

FIGURE 5.40
Pressure distribution in the Paluxy Formation top layer after 500 years – all scenarios
(SI conversion: 1 psi = 6.9 kPa, 1 ft = 0.3 m, 1 D = 9.87e-13 m^2).

As expected, the same behavior was observed in the pressure distribution
profiles. This behavior can be attributed to the fact that, as the permeability
in the confining unit decreases, less pressure dissipation will occur as a
consequence of CO$_2$ migration upwards and out of the Paluxy Formation, so
the post injection pressure distribution will no longer be governed as much
by "fluid loss" as by CO$_2$ plume migration itself. Figure 5.40 summarizes the
results of the simulations for a period of 500 years for all defined scenarios.

5.9 Post Injection Site Care

In November 2010, the Administrator of the Environmental Protection Agency
(EPA) signed the Federal Requirements under the Underground Injection
Control (UIC) for CO$_2$ sequestration wells "Final Rule," as authorized by
the Safe Drinking Water Act (SDWA). The "Final Rule" establishes new
federal requirements for the underground injection of CO$_2$ for the purpose

of long-term underground storage, or geologic sequestration, and a new well class – Class VI – to ensure the protection of underground sources of drinking water (USDWs) from injection-related activities.

EPA's proposed rule making describes the technical criteria for geologic site characterization, fluid movement, area of review (AoR) and corrective action, well construction, operation, mechanical integrity testing, monitoring, well plugging as well as post-injection site care and site closure to protect USDWs.

Post injection site care (PISC) for the Citronelle field CO_2 sequestration project is bounded within the legal framework established by the following regulations:

- Code of Federal Regulations – Title 40 – Protection of Environment – Chapter I, Subchapter D-part 146, section 146.93 "Post-injection site care and closure."
- Alabama Department of Environmental Management Water Division – Water Quality Control Administrative Code 335-6-8-0.25 Class VI Well Post-Injection Site Care and Site Closure Requirements.

Regarding post injection site care and closure, some sections of the regulations establish the following:

- The post-injection site care and site closure plan must include the following information
- The pressure differential between the pre-injection and predicted post-injection pressures in the injection zone(s)
- The predicted position of the CO_2 plume and associated pressure front at site closure as demonstrated in the area of review evaluation required under § 146.84(c)(1)
- A description of post-injection monitoring location, methods, and proposed frequency
- A proposed schedule for submitting post-injection site care monitoring results to the Director pursuant to § 146.91(e)
- The duration of the post-injection site care timeframe and, if approved by the Director, the demonstration of the alternative post-injection site care timeframe that ensures non-endangerment of USDWs.

The main focus of the post injection site care study being performed is to determine the time when a maximum yearly pressure differential has been achieved (stabilization time) based on thresholds of 68.9 kPa (10 psi), 34.5 kPa (5 psi) and 13.8 kPa (2 psi) per year. For this purpose, we monitored yearly pressure data generated in all the simulation runs, for each of the 15,625 grid blocks located in the Paluxy Formation top layer. We also determined the time when the maximum pressure difference threshold was found.

TABLE 5.11

Scenarios Used in PISC

Scenario	Permeability of the Confining Unit (m²)	K_h Range of the Confining Unit (m²–m)	
High permeability	9.87E-16	4.51E-14	7.52E-14
Medium permeability	9.87E-18	4.51E-16	7.52E-16
Low permeability	9.87E-20	4.51E-18	7.52E–18

Based on the findings from the seal quality study, and knowing there will be a greater impact from changes in the permeability of the confining unit, opposed by the effects of the thickness of the seal, a range of "seal conductivity" was defined (permeability × thickness of the confining unit measured in 4.51E-14 m²-m to 7.52E-14 m²-m). The scenarios used for this study are summarized in Table 5.11. The results shown in the following sections will correspond to the pressure difference distribution per time step until the corresponding threshold condition has been satisfied and the distribution profiles of where in the top layer of the Paluxy Formation this condition has been met.

The results for one of the scenarios will be analyzed and discussed here as an example case. After calculating and plotting the yearly pressure differentials we could determine that the all the thresholds are met in the second year after injection (post-injection monitoring) has stopped. The following figures represent the spatial distribution of these pressure differences for 1 year post injection and 2 years post injection for verification purposes.

As observed in Figures 5.41 and 5.42, after the end of the first year post injection, we can still observe large pressure differences, particularly in the surroundings of the injection well. This pressure is dissipated after the second year post injection has elapsed. Comparing them with the three predefined thresholds for pressure variation in yearly bases (68.9 kPa (10 psi), 34.5 kPa (5 psi) and 13.8 kPa (2 psi)), it was possible to determine when and where in the reservoir these conditions were not being satisfied. Figures 5.43 and 5.44 present

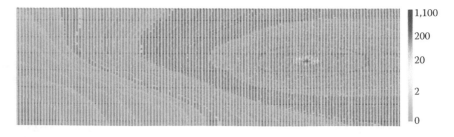

FIGURE 5.41

Yearly pressure difference distribution 1 year post-injection (SI conversion: 1 psi = 6.9 kPa).

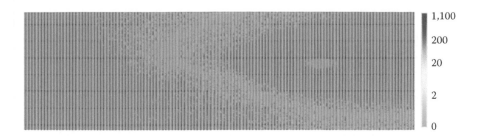

FIGURE 5.42
Yearly pressure difference distribution 2 years post-injection (SI conversion: 1 psi = 6.9 kPa).

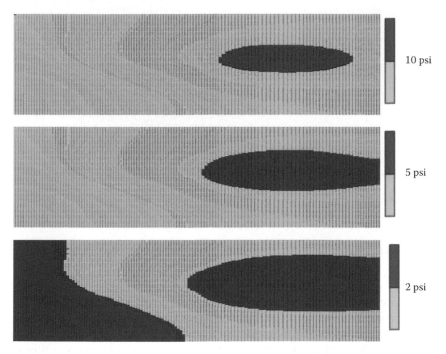

FIGURE 5.43
Threshold verification 1 year post injection (SI conversion: 1 psi = 6.9 kPa).

the pressure difference distributions as before, but this time the formatting was set to determine which parts of the top layer were over or under the pressure thresholds in order to verify the stabilization times after 1 and 2 years of post injection monitoring.

As observed in Figure 5.44, the condition for the three pre-defined thresholds has been fully satisfied after the second year of post injection monitoring, so the pressure is considered to be stabilized under this criterion by the end of the second year after injection has stopped.

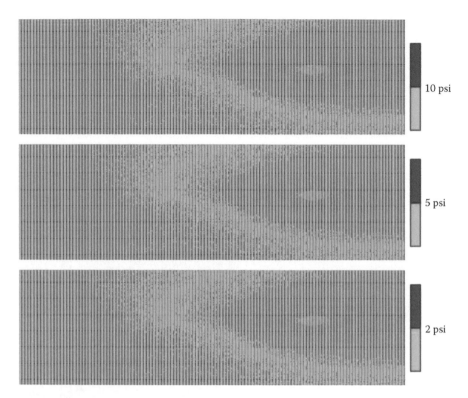

FIGURE 5.44
Threshold verification 2 years post injection (SI conversion: 1 psi = 6.9 kPa).

5.10 The Model's History Match

All the steps for the development of the reservoir simulation model and reservoir study for CO_2 injection into the Citronelle saline aquifer with commercially available software (CMG-GEM) have been explained in Sections 5.5.2 and 5.5.3. Field measurements of CO_2 injection rates were assigned as the operational constraint to the model. In addition to the injection rates, high-frequency, real-time pressure data from two down-hole pressure gauges embedded in an observation well (254 m away from the injection well) is also provided. Several uncertain reservoir properties were tuned within reasonable ranges in order to find a proper match between the simulated pressure results and actual field measurements (53).

Different types of potential risks, like the leakage of CO_2 or brine from the target zone, are generally associated with the geological storage of CO_2. Reservoir simulation and modeling, in addition to the implementation of appropriate monitoring techniques, are considered to be expedient tools for CO_2 risk management.

Reservoir pressure/temperature measurement by down-hole gauges has been widely used in the oil and gas industry for reservoir monitoring, well test analysis, and history matching. In CO_2 sequestration projects, real-time reservoir pressure measurements can provide indications of CO_2 migration/leakage. Meckel and Hovorka (54) interpreted permanent down-hole gauges (PDGs) data collected from a single well at injection and above the zone monitoring interval for CO_2 injection at Canfield. They suggested almost no inter-formational communication (vertical) at the site (based on analysis of pressure changes) due to seven injection and nine production wells' activities. Tao et al. (55) analyzed the same pressure and temperature data (collected from the monitoring well at Cranfield) and concluded a very small leakage had occurred from the injection interval to the overlying formation. PDG data can also provide valuable information for reservoir simulation models.

Reservoir models can be used for the assessment of CO_2 storage capacity, well injectivity, CO_2 trapping mechanisms, CO_2 plume extension, and reservoir pressure build up. Sifuentes et al. (45) studied the effect of different physical parameters on CO_2 trapping in Stuggart Formation in Germany. In order to determine the contribution of each parameter to CO_2 trapping, they used reservoir simulation coupled with experimental design to perform a sensitivity analysis. Torn et al. (56) carried out almost the same sensitivity analysis on Mt. Simon sandstone model to assess storage capacity and safety issues. Senel et al. (57) performed a reservoir simulation and uncertainty analysis study on CO_2 injection in the same formation (Mt. Simon sandstone, USA) incorporating more geophysical and petro-physical data. They investigated the effect of uncertainty on trapping mechanisms and the CO_2 area of extension by providing probabilistic estimates. Masoudi et al. (58) coupled a geo-mechanical and a simulation model in order to study the feasibility and risks associated with CO_2 injection in M4 Field (East Malaysia). They determined the maximum allowed reservoir pressure by considering the cap-rock integrity for different CO_2 injection scenarios.

The reservoir simulation performance must be validated by checking if the model is able to regenerate the past behavior of a reservoir. History matching of oil and gas reservoir models is more often applied than history matching of CO_2 storage models due to the availability of large amounts of production and/or injection data. For CO_2 storage projects, especially those in saline formations, reservoir history data is limited to the injection rate and down-hole injection/observation well pressure. Mantilla et al. (59) used probabilistic history matching software known as Pro-HMS, which incorporated injection data from active injection and inactive observation wells. They implemented Pro-HMS to a synthetic model – CO_2 storage in an aquifer with one/three injection and one observation wells – to obtain high-permeability streaks by using only injection and pressure data. In another history matching attempt, Krause et al. (60) conducted core flooding (brine/CO_2) followed by numerical simulation of the experiment. They matched the simulation results with experimental data by calculating the permeability, using porosity and capillary pressure data. Xiao et al. (61) studied the numerical simulation of CO_2/EOR and storage in a pilot-5spot pattern unit of SACROC field. Since the target storage field had a long-term production/injection history, they performed history matching for the gas, oil and water production of five wells. They also predicted the reservoir performance for three EOR injection schemes and analyzed the CO_2 storage capacity for different CO_2 trapping mechanisms.

This chapter explains one of the very few examples of this kind that aims to history match a reservoir simulation model of CO_2 injection in Citronelle Dome (saline formation). The available field data for history matching comprises ten months of CO_2 injection rates as well as pressure data coming from two gauges installed in the observation well.

The locations of the injection and observation wells are shown in Figure 5.45. CO_2 injection started on August 20, 2012 at a rate of 26×10^3 m^3/day (918 Mcf/day). After that time, the injection rate increased with an oscillating trend (because of operative difficulties) until the end of September 2012 when it reached 25.4×10^4 m^3/day (9 MMcf/day, or the targeted rate). The injection then continued until August 2013 with a stable rate, although periodic shut downs occurred. The daily injection rate from the beginning up to August 2013 is shown in Figure 5.46. These injection rates were used in the reservoir simulation model as operational constraints.

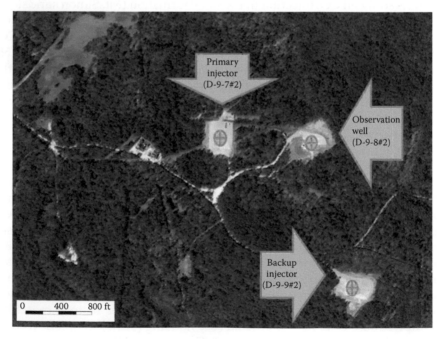

FIGURE 5.45
Locations of the CO_2 injection, observation, and backup injection wells in Citronelle Dome.

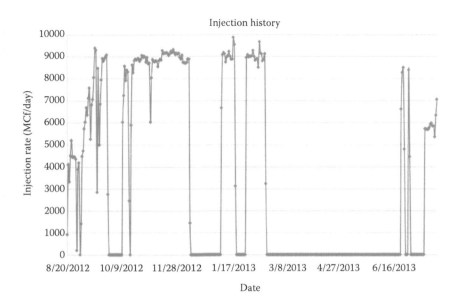

FIGURE 5.46
CO_2 injection rate history (SI conversion: 1 ft³ = 0.028 m³).

In the observation well (D-9-8#2) at Citronelle field (Figure 5.45), two PDGs (PDG-5108/5109) were installed at different depths of 2,870 and 2,878 m true vertical depth (TVD) in order to provide real-time pressure and temperature readings during and after the injection period. The pressure data is available from mid-August of 2012 until August 1, 2013, is recorded for every minute, and lists 1,440 records daily. There are some gaps in the pressure records due to an on-site computer failure. Since history matching the data on a minute basis was computationally expensive and time consuming, the pressure data was summarized by averaging over each day. The results of the actual pressure data on a daily basis are illustrated in Figure 5.47.

Initially, a base case reservoir model was developed considering the reservoir properties summarized in Table 5.12. Porosity maps for each simulation layer were acquired by interpreting 40 well logs. In this model, the operational constraints were the actual CO_2 injection rates and the maximum bottom-hole pressure limit of 43,437 kPa (6,300 psi). The solubility of CO_2 in the brine was not considered in the base model. Block pressures for the grids corresponding to the PDGs were compared with the actual data. The simulated pressure data using the base model is plotted against the actual pressure history in Figure 5.48.

The simulation data do not match either at the start point or in terms of the difference between the values of the two gauges. The initial reservoir pressure was adjusted by changing the reference pressure to 30,130 kPa (4,370 psi) at the datum depth of 2,870 m. The pressure gradient between the PDGs was 14 kPa/m and between the simulation grids was 9.73 kPa/m. Therefore, it was concluded that the brine density should be set at a higher value in order to

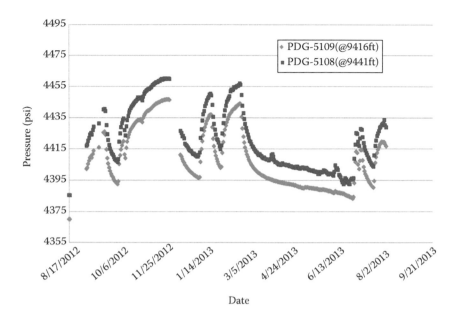

FIGURE 5.47
Daily pressure data from PDGs at the observation well (SI conversion: 1 psi = 6.9 kPa).

mimic the same pressure gradient so that the brine density at the reservoir conditions can be calculated using the following equation:

$$\tilde{\rho}_{br} = \tilde{\rho}^0_{br}\left[1 + c_{br}(p - p^0)\right]$$

Keeping the brine compressibility unchanged, the density of brine should be altered to 1,393.6 kg/m^3. As mentioned in Sections 5.2 and 5.3, a thorough sensitivity analysis was performed to study the effect of several uncertain reservoir parameters on the pressure behavior in the observation well. The results of the sensitivity analysis showed that permeability significantly contributed to injectivity, CO$_2$ plume extension, and reservoir pressure. Using the available core data (not taken from Citronelle field) a porosity–permeability

TABLE 5.12

Reservoir Parameters and Properties (Base Model)

Parameter	Value	Parameter	Value
Permeability (m^2)	4.50E-13	Water density (kg/m^3)	993
Temperature (K)	383	Water viscosity (cP)	0.26
Salinity (ppm)	100,000	Water compressibility (1/Pa)	4.60E-10
Residual gas saturation	0.35	K_v/K_h (permeability ratio)	0.1

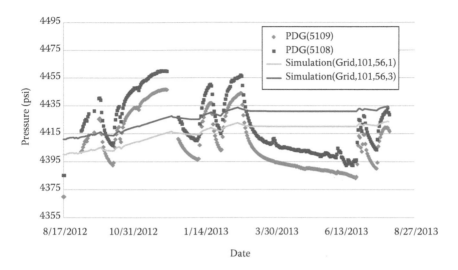

FIGURE 5.48
Actual pressure verses simulated pressure in the base model (SI conversion: 1 psi = 6.9 kPa).

cross plot was generated for the Paluxy Formation (Figure 5.18). Available data from core samples taken from the injection well demonstrated the dominance of the conductive rock type in the vicinity of the injection area (42). Also, the vertical to horizontal permeability ratio was calculated to be 0.58 using core data analysis. Modification of the pressure reference, brine density, and permeability, in addition to setting zero transmissibility between the sand and shale layers, resulted in the pressure predictions illustrated in Figure 5.49.

Through the implementation of modified parameters in the model, the prediction results for initial pressure and pressure gradient were similar to the actual data. However, the model pressure predictions did not follow the PDG pressure trend correctly. As shown in Figure 5.49, the reservoir simulation results underestimated the actual data during the first four months after injection, and overestimated the rest of the pressure history. Additionally, simulation pressure drawdowns reached a stable trend much more quickly than the actual data. This behavior can be explained by the fact that higher permeability (in the model) resulted in a lowering of the time for the pressure drawdown to reach a steady trend. Therefore, it was necessary to decrease the permeability in the model to adjust the pressure drawdown behavior. On the other hand, lowering the permeability led to a reduction in CO_2 injectivity. As a result, the reservoir model was divided into two regions: (a) grids in the vicinity of the injection zone (20 × 20 grids around the injection well) and (b) grids outside the injection zone (Figure 5.50). To correct the model's pressure drawdown trend, a dual modification in the reservoir permeability was carried out by increasing the permeability in region (a) and decreasing the permeability in region (b).

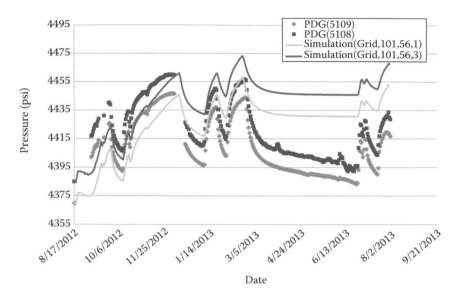

FIGURE 5.49
Model's pressure results and actual history: modified pressure reference, brine density, and permeability (SI conversion: 1 psi = 6.9 kPa).

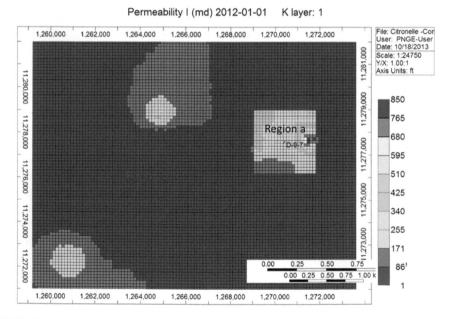

FIGURE 5.50
Two permeability regions in the reservoir (SI conversion: 1 ft = 0.3 m, 1 D = 9.87e-13 m²).

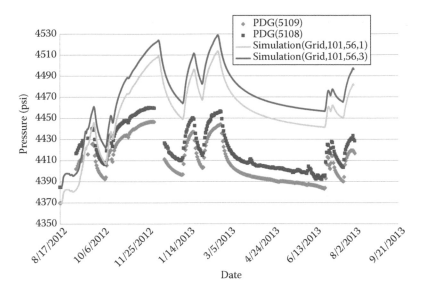

FIGURE 5.51
Model's pressure results and actual history: modified permeability in different reservoir regions (SI conversion: 1 psi = 6.9 kPa).

As shown in Figure 5.51, although modifications in the model's permeability improved the pressure drawdown behavior, pressure predictions were overestimated considerably compared with the actual PDG data. To lower the pressure results, the solubility of CO_2 in the brine (aqueous phase) was incorporated into the model (51). More importantly, a volume modifier was assigned to the grids at the east boundary of the reservoir (Figure 5.52). This accounted for the fact that the reservoir boundary and volume might be bigger than assigned to the model. To develop the geological model, the top and thickness of the sand layers were picked for the log data of 14 wells crossing at the injection well and then correlated (Figure 5.52).

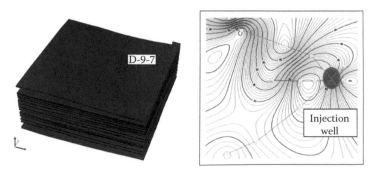

FIGURE 5.52
Increased volume modifier at the east boundary. Right: well cross sections.

Due to the limited amount of information (just two well logs) at the east side of the injection well, it was not possible to estimate the extension of the sand layer on that area. It was therefore probable that more reservoir volume existed outside the boundary of the geological model. Adding more volume to the reservoir (sand layers) resulted in lowering of the pressure prediction. After activating the CO$_2$ solubility in the brine phase and tuning the "volume modifier," a good match between the model results and actual pressure data, with less than 0.001% average error, was achieved (Figure 5.53).

5.10.1 Model Validation

The history-matched model showed very good precision in generating ten months of pressure results that resembled the actual field measurements. To study the predictability of the reservoir model, the last three months (August 1 to October 30, 2013) of actual injection/pressures were not used in the history matching and were set aside for forecast validation (Figure 5.54). During these three months, CO$_2$ was injected steadily according to the targeted rate 26.8 × 10^4 m^3/day (9.48 MMcf/day). Injection was subject to a few shut downs, resulting in an average rate of 22.6 × 10^4 m^3/day (7.98 Bcf/day). Consequently, the reservoir pressure increased during August 2013, followed by some drawdowns due to no injection in September 2013 and gentle build up during the last month.

Considering the last three months of the injection rate profile as the model's operational constraints, simulation pressure predictions were obtained. The pressure prediction results are plotted versus actual data in Figure 5.55. The

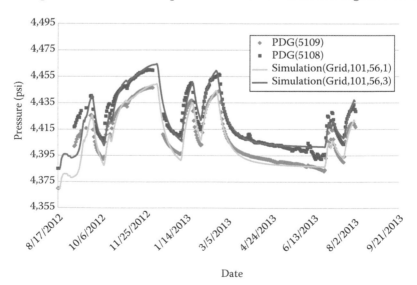

FIGURE 5.53
Model's pressure results and actual history: final history match (SI conversion: 1 psi = 6.9 kPa).

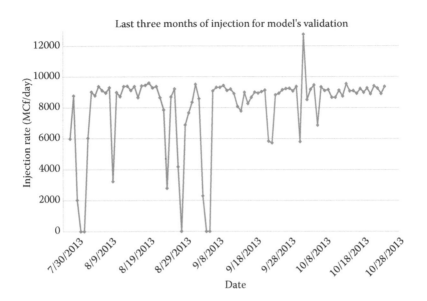

FIGURE 5.54
Last three months of injection rate (SI conversion: 1 ft³ = 0.028 m³).

prediction has precisely captured the actual data trend such that the average errors for gauges 5109 and 5108 are 0.12% and 0.073%, respectively, which is quite satisfactory.

Ten months of CO_2 injection in the Citronelle Dome's saline formation (Paluxy) was modeled by numerical reservoir simulation. The presence of

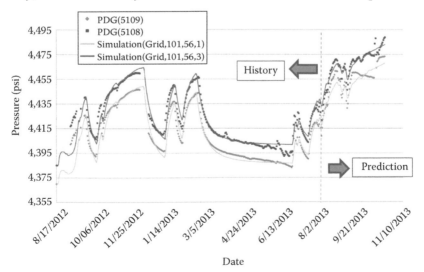

FIGURE 5.55
Model's pressure result and actual data for prediction and history (SI conversion: 1 psi = 6.9 kPa).

two PDGs at the observation (monitoring) well was considered in the model. A comprehensive sensitivity analysis was performed to assess the effect of the uncertainty of several reservoir parameters on the model's pressure (at the observation well) behavior. The analysis was used to match the history of the actual field pressure data with the model's prediction by tuning the reservoir parameters. By modifying the brine density, the permeability in two reservoir regions, the vertical to horizontal permeability ratio, the CO_2 solubility in brine, and the reservoir volume, a reasonable match (less than 0.001% error) between the actual and model's pressure data was achieved. This model was validated using the last three months of pressure-injection profiles and showed acceptable predictability. However, history matching of the numerical model is a non-unique solution to a complex problem, and other combinations of reservoir parameter modification may possibly result in the same match between the actual data and model's data.

6

CO₂ *Storage in Shale Using Smart Proxy*

Shahab D. Mohaghegh and Amirmasoud Kalantari-Dahaghi

CONTENTS

Continuing the development of organic-rich and extremely low-permeability shale reservoirs in the United States has the potential to positively impact the future of carbon storage. Due to the unique characteristics of shale reservoirs, not only can CO_2 be safely stored, it can also be preferentially adsorbed and displace methane, leading to enhanced gas recovery (EGR).

However, CO_2 storage in depleted or nearly depleted shale formations is not completely risk free. Thus, prior to making an economic commitment to a full-field CO_2 storage project in shale, a systematic analysis of the complete set of variables must be considered during the planning process. Numerical modeling and simulation has been used as a tool to provide insight into how the system may operate in order to further understand the feasibility of projects as well as assist in their design and operation, and to predict any changes that may occur.

In order to perform a comprehensive uncertainty analysis, a large number of simulation runs are required. Designing and running simulation cases to model enhanced gas recovery and storage (EGR&S) in shale by applying the explicit hydraulic fracture (EHF) modeling technique is long and laborious, and its implementation is computationally expensive.

This chapter introduces a data-driven approach with pattern recognition algorithms to develop a new generation of shale smart proxy models at the hydraulic fracture cluster level, in order to replicate the results of reservoir simulation models. For a more accurate analysis, instead of the commonly used mechanistic models, a history-matched hydraulic fractured Marcellus shale pad with multiple stages/clusters is used as a base case.

The detailed procedure for development (training and calibration) of the data-driven smart proxy model is explained here and then validated using blind simulation runs. Upon completion and validation, the developed data-driven smart proxy model is capable of accurately reproducing the calculated CO$_2$ injection, CO$_2$/CH$_4$ production profiles, and CO$_2$ breakthrough time from the numerical simulation model, for each cluster/stage and horizontal lateral. Joint use of the deterministic reservoir model with the data-driven smart proxy model can serve as a novel screening and optimization tool for techno-economic evaluation of the CO$_2$-EGR&S process in shale systems.

6.1 Using Numerical Simulation for Shale

Using numerical reservoir simulation to model fluid flow in the presence of extensive natural fracture networks that are stimulated and activated through a large number of hydraulic fractures, which is designed through multiple clusters and stages, and the subsequent production of hydrocarbon from these lateral wells in shale, is an interesting subject for discussion. The view of one of this chapter's authors on this topic has been extensively published (62) and discussed in previous publications (63). However, regardless of this view of the technical accuracy and correctness of using numerical reservoir simulation in modeling storage and transport phenomena in shale, the fact remains that many in our industry are still using this technology to perform analyses and make decisions. Therefore, this chapter is dedicated to the use and application of big data analytics and data-driven solutions to ease the burden of performing such analyses, and making decision making a practical reality. This chapter is a demonstration of the author's intimate involvement with the use of numerical reservoir simulation to model fluid flow in shale.

6.2 Challenges and Solutions of the Numerical Simulation of Shale

Reservoir simulation is used as a tool for the detailed analysis of production from shale and the CO$_2$-EGR&S process. The base simulation model, which is properly calibrated with the available production history in the primary recovery phase, should be used to model the reservoir response during the CO$_2$ injection process (64).

Integration of natural fracture networks and hydraulic fractures with desorption–diffusion phenomena in the shale matrix is a challenging task in the simulation of shale gas reservoirs. EHF and stimulated reservoir volume (SRV) techniques are the two most commonly used techniques for the simulation of shale gas production. EHF modeling attempts to predict the impact of hydraulic fracturing at the cluster/stage level by incorporating hydraulic fracture characteristics into the simulation model. Model setup for the EHF technique is long and laborious, and its implementation is computationally expensive, so much so that it becomes impractical to model and history match beyond a single pad.

Furthermore, to model the CO_2-EGR&S process, a compositional simulator should be utilized, which significantly increases computation time. Therefore, the perennial challenge in modeling primary production and the CO_2-EGR&S process is to strike a balance between the explicit representation of reservoir complexity and long simulation run times for multiple realizations. Smart proxy models can be developed and used to overcome the aforementioned problems and to address the under-utilization of numerical simulation models (65).

6.3 Traditional vs. Smart Proxy Models

Traditional proxy models are either mathematically driven simplified models or statistically generated response surfaces that are developed to substitute large numerical reservoir simulation models by replicating the simulation model's output and provide fast approximated solutions to be used efficiently in development planning, uncertainty analysis, and operational design optimization.

The most frequently used traditional proxy models in the oil and gas industry are reduced order models and response surfaces, which reduce the simulation run time by approximating the problem and/or the solution space (66). Response surfaces are categorized as statistical-based proxy models and require a large number of simulation runs to facilitate optimization and uncertainty analysis.

According to Mohaghegh (67), there are two main problems associated with statistical approaches when they are employed to build proxy models (response surfaces), especially when applied to problems with well-defined physics behind them, such as numerical reservoir simulation models: (1) the issue of "correlation vs. causation" and (2) imposing a pre-defined functional form such as linear, polynomial, exponential, etc., to the data being analyzed or modeled. These approaches will fail when data representing the nature of a given complex problem does not lend itself to a pre-determined functional form and changes behavior frequently.

Unlike traditional proxy models, data-driven smart proxy models take a different approach to developing proxy models. Unlike reduced order models, the physics and space–time resolution are not reduced, and instead of using the pre-defined functional forms that are more frequently used to develop response surfaces, a series of machine learning algorithms that conform to system theory are used for training, with the ultimate goal of more accurately modeling the intricacies of a developed shale CO_2-EGR&S numerical reservoir simulation model. Developed in 2005, smart proxy modeling technology was first introduced to the petroleum industry in 2006 (20–22). The technology was then used in multiple projects and studies, including the one presented in this chapter.

An ensemble of multiple, interconnected, adaptive neuro-fuzzy systems create the core for development of the smart proxy models. An artificial neural network is an information-processing system with certain performance features analogous to biological neural networks. In this system, neurons are clustered into different layers, including input, output, and hidden layers. The number of parameters in the dataset defines the number of neurons in the input layer. One or multiple output(s) can be defined during neuro-model development. One important step toward building a smart proxy model with machine learning is feature extraction by including the hidden layer(s) and defining the hidden neurons in that layer, which increases the dimensionality that accommodates classification and pattern recognition (68). Neural networks can be divided into two main classes based on training algorithms: unsupervised and supervised. In the supervised training mode, both input and output in the form of a spatio-temporal database are presented to the neural network to permit learning on a feedback basis. Different clustering algorithms can be used to partition the cases into three main groups: training, calibration, and validation (68).

During the training process, the weights between the processing elements are adjusted. Memorization, also known as over-training, is an issue that must be avoided during the development of a data-driven model. The calibration portion of the data is used for that purpose, to continuously examine the trained neural network's generalization capabilities and trigger a halt to the training once an acceptable model has converged. The smart proxy model development workflow is completed once the verification dataset is used to test the predictive capability of the model.

6.4 History Matching Production from the Marcellus Shale

In order to make a realistic model, information from 77 Marcellus shale gas wells in Southwestern Pennsylvania – with a total number of 652 stages of hydraulic fracture and 1,893 clusters in an area of about 53,241 acres – was used to develop a numerical reservoir simulation model. A representative pad (the WVU pad) – with six horizontal laterals in terms of reservoir and hydraulic fracture characteristics in the study area – was selected to perform the numerical reservoir simulation. A compositional, matrix-discretized, dual-porosity reservoir simulation model consisting of 200,000 grid blocks was constructed. Hydraulic fractures were explicitly included in the three-layer simulation model. The resulting numerical reservoir simulation model included nine simulation layers in the refined regions. The simulation model was used to history match methane production from a pad with six lateral wells. For simplicity, this pad will be referred to as the WVU pad throughout this chapter. The history-matched model was then used to forecast production from the reservoir.

Three years of production from six laterals in the WVU pad were successfully history matched, and production was forecast for 90 years to generate a base case for a further shale CO_2-EGR&S study. Figure 6.1 shows the entire study area, with 77 horizontal wells and the location of the target WVU pad used in this study for numerical simulation of CO_2 injection.

Simulation of the CO_2 injection process in shale was started by defining and running a large number of injection scenarios to recommend a proper set of injection realizations for CO_2-EGR&S. To optimize the process, a function with multiple simultaneous objectives (a multi-objective function) was defined. The objectives of the function were (1) to maximize CH_4 recovery, (2) to delay the CO_2 breakthrough time, and (3) to maximize the volume of stored CO_2 in the reservoir.

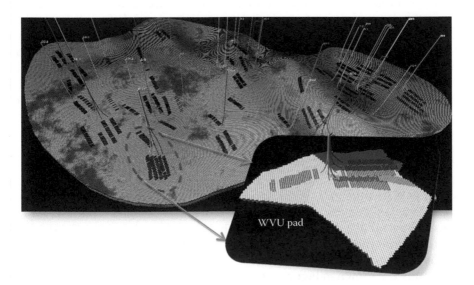

FIGURE 6.1
Shale numerical reservoir simulation model (3D) and location of the targeted WVU pad.

As stated already, designing and running simulation cases to model EGR&S in shale by applying the EHF modeling technique is a long and laborious process, and its implementation is computationally expensive, thus requiring the development of a proxy model as an alternative for fast-track uncertainty analysis. Development of a smart proxy model was selected for this purpose.

6.5 Generating the Spatio-Temporal Database

The most important step in the development of a smart proxy for a reservoir model is assimilation of a comprehensive spatio-temporal database by considering the uncertainty domain and operational limitations, which form the foundation of a data-driven smart proxy model for the CO$_2$-EGR&S process. The success of the data-driven smart proxy model depends greatly on the degree to which the training dataset represents the fluid flow and storage behavior of the shale reservoir.

In this study, the CO$_2$ injection plan included injecting through four patterns with different matrix, fracture characteristics, and spacing between injector and producer. From a practical point of view, CH$_4$ production from the prospective injection well should reach an economic limit before it is converted into an injector well. To meet that criterion, when the production well has drained 75% of its recoverable reserve (based on 100 years EUR [estimated ultimate recovery]), CO$_2$ injection starts and continues until the end of $t = 100$ years.

Five simulation runs were defined for each case (a total of 20 runs) to cover the desired range of reservoir characteristics and different operational conditions. The generated comprehensive spatio-temporal database was used to train, calibrate, and validate a multilayer feed-forward, back-propagation neural network to accurately mimic the reservoir simulator behavior, to re-generate CH$_4$ production and CO$_2$ injection profiles, and also to predict the CO$_2$ breakthrough time and CO$_2$ production profile for each cluster of hydraulic fractures in the producer and injector(s), as well as corresponding laterals. The summation of gas rates from all clusters generates a production/injection profile for the corresponding horizontal lateral. This process is demonstrated in Figure 6.2.

In order to take the validation one step further, the developed data-driven smart proxies were validated with a set of completely blind simulation runs that were not used during the training process. A detailed workflow for the development of the data-driven smart shale proxy model for the CO$_2$-EGR&S process is presented in Figure 6.3.

The four production cases of injection/production patterns considered in this study are shown in Figure 6.4. As mentioned previously, for each pattern (or case) shown in this figure, five simulation runs are defined to cover the desired range of reservoir characteristics (i.e., matrix, natural fracture and

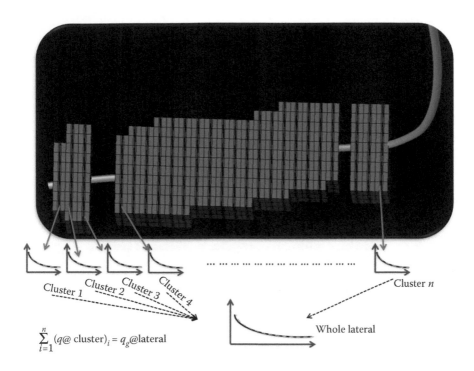

FIGURE 6.2
Illustration of a production profile generated by a data-driven smart proxy model at hydraulic fracture cluster level.

hydraulic fracture properties, as well as different Langmuir isotherms) and different operational constraints (flowing bottom-hole pressure and bottom-hole injection pressure).

In the first three proposed simulation cases, the average distance between a producer–injector pair varies from 275 to 616 m. In Case 4, the hydraulic fractures for both production and injection laterals are overlapping and therefore early breakthrough can be expected for this pattern. This case is included in the spatio-temporal database to make sure that the developed data-driven smart proxy model has the opportunity to see such a case and learn its characteristics, so that it will be robust enough to predict production and injection performance in any situation. In order to take into account the impact of the properties of different grid blocks, on each cluster a production/injection "tier system" was defined. Three different tiers of grid blocks surrounding the modeled clusters of hydraulic fractures were defined, and the corresponding property for each tier was calculated by averaging the properties of all the grid blocks in the corresponding tier. Moreover, to train the neural networks on the potential interference effect between clusters, they were divided into four classes based on their relative locations with respect to each other.

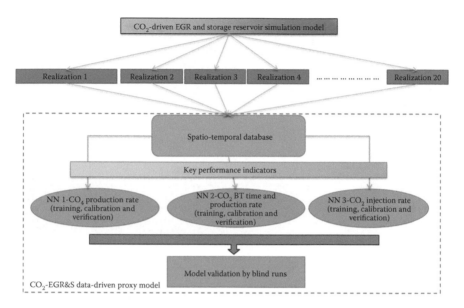

FIGURE 6.3

CO$_2$-EGR&S data-driven smart shale proxy model development workflow.

Table 6.1 lists the required information needed to build the training, calibration, and validation dataset to be used for development of the smart proxy model.

The back-propagation algorithm was used to perform training of the neural networks involved in the study. Different clustering algorithms can be used to partition the cases into three main groups: training, calibration, and validation. During the training process, the weights between the processing elements were properly adjusted. The proxy model development workflow was complete once the neural networks reached the best calibration and the trained model passed the final predictive competence test.

6.6 Results and Discussion

In this section, the results for four data-driven CO$_2$-EGR&S smart proxy models at hydraulic fracture cluster level are presented. Data-driven smart proxies are developed to predict (1) CH$_4$ production rate, (2) CO$_2$ injection rate, (3) CO$_2$ breakthrough time, and (4) CO$_2$ production rate for each hydraulic fracture cluster. When the results from all clusters are aggregated, the result is the corresponding value for the entire lateral. The process is repeated for each lateral. In the next several sections we present detailed results from a commercial numerical reservoir simulator and compare them with the results

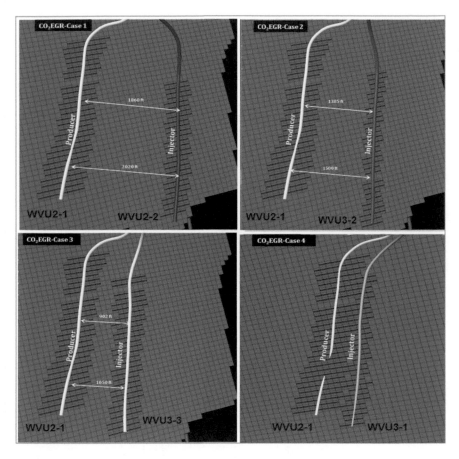

FIGURE 6.4
CO$_2$-EGR&S process simulation cases (producer and injector patterns) (902–2,020 ft is equivalent to 275–616 m).

generated by the smart proxy developed during this study. It is important to note that while a single run of the numerical reservoir simulator takes tens of hours to complete, a single run by the smart proxy model takes only a few seconds.

6.6.1 Prediction of CH$_4$ Production Profiles

The first data-driven CO$_2$-EGR&S smart proxy model was developed to generate an annual CH$_4$ production rate for each cluster of hydraulic fractures for 100 years; this had previously been generated by the commercial reservoir simulator. The 20 simulation runs that were designed and executed based on four injector/producer patterns had generated 1,160 unique CH$_4$ production profiles with different petro-physical characteristics for each grid block (and therefore for each tier system), natural and hydraulic fracture characteristics,

TABLE 6.1

Required Information for CO$_2$-EGR&S Data-Driven Smart Shale Proxy Models Development

Matrix porosity (0.054–0.125)	Matrix permeability (0.0001–0.0008 mD)	Natural fracture porosity (0.01–0.035)	Natural fracture permeability (0.001–0.004 mD)	Sigma factor (0.005–0.08)
Hydraulic fracture height (30.48–38.1 m)	Hydraulic fracture length (60.96–335 m)	Hydraulic fracture conductivity (0.031–1.65 mD-m)	Net pay thickness (34.44–39 m)	CH$_4$-Langmuir volume (1.716–2.84 m^3/ton)
CH$_4$-Langmuir pressure (4,137–5,447 kPa)	CO$_2$-Langmuir volume (2.18–3.74 m^3/ton)	CO$_2$-Langmuir pressure (2,758–4,000 kPa)	CH$_4$-diffusion coefficient (0.18–0.37 m^2/day)	CO$_2$-diffusion coefficient (0.092–1.86 m^2/day)
Bottom-hole injection pressure (11,583–23,167 kPa)	Flowing bottom-hole pressure (896–5,654 kPa)			

sorption features (sorption time, diffusion coefficients, and Langmuir isotherms), and operational constraints.

Before starting the training, calibration, and validation process, the fuzzy pattern recognition (69) technique was used to determine the relative importance and the impact of the input parameters on CH_4 production. According to this analysis, the first set of the most influential parameters on CH_4 production are duration of production, operational constraints, producer/ injector patterns, and the relative location of the clusters. The second set of the most influential parameters are those that control the reserve, such as isotherms, net pay thickness, natural fracture porosity, and matrix porosity.

Some other parameters – such as hydraulic fracture components, natural fracture properties that control the accessibility of gas to the clusters – can be considered a third group of parameters that have significant influence on production. They are essential to the productivity of shale gas wells at the beginning of production, but for long-term production performance, operational constraints and parameters that control the reserve play a more important role.

To continue with the training and validation procedure, the neural network architecture was designed to have one intermediate hidden layer with 55 hidden neurons, which were selected based on the number of data records available and the number of input parameters selected in each training process. In the training process, the dataset was partitioned into three separate segments by using the intelligent partitioning technique (69). The data was partitioned with 70% of data to be used for training, 20% for calibration, and 10% for verification.

The training, calibration, and verification results for annual CH_4 production rate (Mscf/year) (from left to right) are shown in Figure 6.5. In these plots, the x axis corresponds to the neural network predicted gas rate, and the y axis shows the gas rate simulated by the commercial reservoir simulator.

The result, with an R^2 of more than 0.99 in all steps, shows the successful development of a data-driven smart shale proxy model for CH_4 production profile prediction.

Figures 6.6 and 6.7 show some examples of a comparison of reservoir simulation output for the annual CH_4 production rate during the CO_2-EGR&S process with the predicted CH_4 production rates from the data-driven smart proxy model for some of the clusters and some of the laterals for all four production and injection patterns/cases.

In all the plots, blue dots represent the annual CH_4 production rate generated by the commercial numerical reservoir simulator, and the solid red line is the results of the data-driven CO_2-EGR&S smart proxy model. The results are self-descriptive enough to show the capability of the data-driven smart proxy model in predicting the CH_4 production profile at a hydraulic fracture cluster and lateral level for different production/injection patterns.

In all four cases, WVU2-1 (the left lateral in Figures 6.6 and 6.7) continues CH_4 production for 100 years, while CH_4 production from WVU2-2 (Case 2),

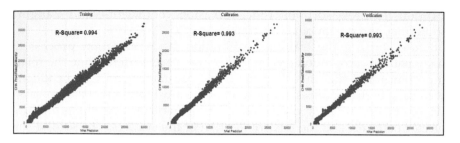

FIGURE 6.5

CH$_4$ production rate (Mscf/year): training, calibration, and verification results (from left to right) (0–30,000 Mscf/year is equivalent to 0–849,600 m^3/year).

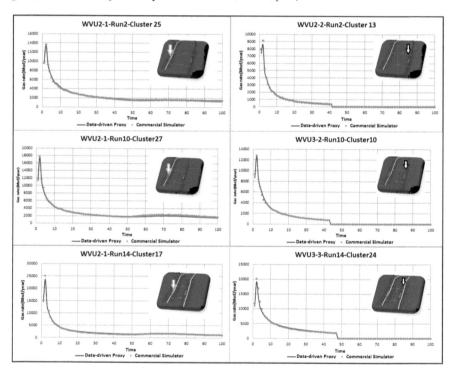

FIGURE 6.6

Comparison of CH$_4$ production rates (Mscf/year) from the simulator and the CO$_2$-EGR&S data-driven smart proxy model for some of the clusters (Cases 1, 2, and 3) (0–30,000 Mscf/year is equivalent to 0–849,600 m^3/year).

WVU3-2 (Case 2), WVU3-3 (Case 3), and WVU3-1 (Case 4) is stopped after producing 75% of the accessible gas, and they are then converted to CO$_2$ injection wells. A boost in the CH$_4$ production profile in WVU2-1 well can be observed when the CH$_4$ displaced by CO$_2$ (during the counter diffusion process) reaches the producing well after 8–20 years, depending on the distance, reservoir characteristics, and bottom-hole injection pressure.

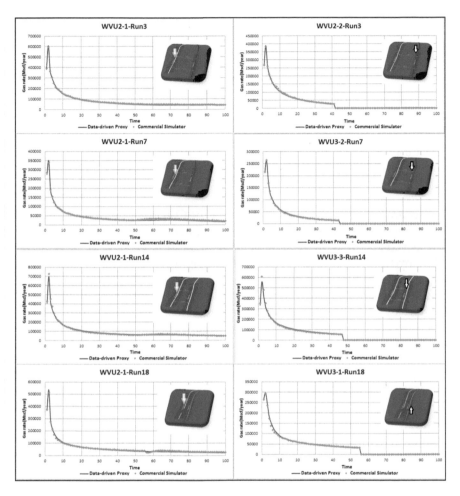

FIGURE 6.7
Comparison of CH_4 production rates (Mscf/year) from the simulator and CO_2-EGR&S data-driven smart proxy model for some of the laterals (Cases 1, 2, 3, and 4) (0–800,000 Mscf/year is equivalent to 0–2.27E+07 m³/year).

6.6.2 Prediction of CO_2 Injection Profiles

The second data-driven CO_2-EGR&S smart proxy model was developed to mimic the numerical reservoir simulation model's behavior and replicate annual CO_2 injection rates (Mscf/year) from the injection startup date until $t = 100$ years. The same procedure as used to build the first smart proxy model for CH_4 production was followed.

The representative database, with 116,000 pairs of input–output data, is used to identify the critical parameters and their degree of influence on CO_2 injectivity in the shale formation. The starting time of injection, operational constraints, and injection/production configuration and relative location of

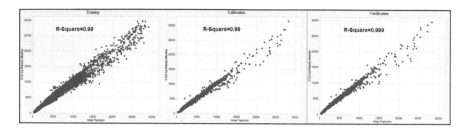

FIGURE 6.8
CO$_2$ injection rate (Mscf/year): training, calibration, and verification results (from left to right) (0–30,000 Mscf/year is equivalent to 0–849,600 m^3/year).

clusters are top ranked parameters during the CO$_2$ injection process, and CO$_2$/ CH$_4$ Langmuir isotherms, fracture porosity, net pay, and fracture half-length are the next top influential parameters for long-term CO$_2$ injection (44–58 years).

The hydraulic fracture conductivity being a low-ranked parameter does not necessarily diminish its importance during the CO$_2$ injection process. It should be noted that hydraulic fracture properties are critical components in allowing CO$_2$ injection in nano-darcy-permeability shale to be initiated. When injection is started and lasts for many years, other parameters come into the picture and show their contribution over the long term.

By having a better understanding of the key parameters during the CO$_2$ injection process, the spatio-temporal database is partitioned as 70% training and 30% blind data (20% calibration and 10% validation). The designed neural network for predicting the CO$_2$ injection profile has one hidden layers with 58 hidden neurons.

Cross plots for the values of CO$_2$ injection rate (Mscf/year) predicted by the neural network vs. the numerical reservoir simulation model's results for training, calibration, and verification steps (from left to right) are shown in Figure 6.8. In these plots, the x axis corresponds to the neural network predicted CO$_2$ injection rate (Mscf/year) and the y axis shows the results from the numerical simulation model for the CO$_2$ injection rates.

The data-driven CO$_2$-EGR&S smart proxy model has been successfully developed to predict CO$_2$ injection rates (Mscf/year) with an R^2 of 0.99 in all steps (training, calibration, and verification). Figures 6.9 and 6.10 show some examples of the comparison of reservoir simulation output for annual CO$_2$ injection rates during the CO$_2$-EGR&S process compared with the predicted annual CO$_2$ injection rates generated by the data-driven smart proxy model for some of the clusters and for the laterals.

In all the plots, blue dots represent the CO$_2$ injection rate (Mscf/year) generated by the commercial numerical reservoir simulation model, and the solid red line is the result of the data-driven CO$_2$-EGR&S smart proxy model. The results show that the data-driven smart proxy model predicts the CO$_2$ injection profile at the hydraulic fracture cluster and full lateral levels for different production/injection patterns with high accuracy.

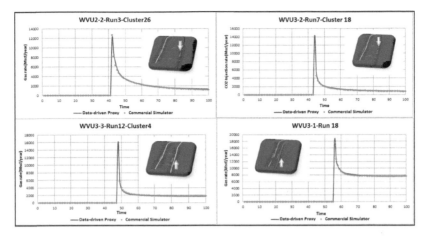

FIGURE 6.9
Comparison of CO_2 injection rates (Mscf/year) from the simulator and CO_2-EGR&S data-driven smart proxy model for some of the clusters (Cases 1, 2, 3, and 4) (0–20,000 Mscf/year is equivalent to 0–566,400 m³/year).

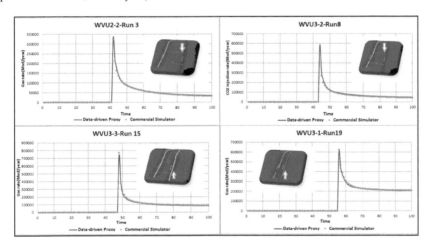

FIGURE 6.10
Comparison of CO_2 injection rates (Mscf/year) from the simulator and CO_2-EGR&S data-driven smart proxy model for some of the laterals (Cases 1, 2, 3, and 4) (0–800,000 Mscf/year is equivalent to 0–2.55E+07 m³/year).

6.6.3 Prediction of CO_2 Breakthrough Time and Production Profiles

The final data-driven CO_2-EGR&S smart proxy model was developed to replicate the CO_2 production profiles (Mscf/year) for offset producing wells. CO_2 breakthrough occurrence is a function of the reservoir characteristics, hydraulic fracture characteristics, sorption features, and bottom-hole injection pressure. The same procedure for building the first and second smart proxy models (CH_4 production and CO_2 injection profiles) is also followed here.

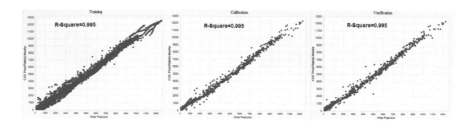

FIGURE 6.11
CO$_2$ production rate (Mscf/year): training, calibration, and verification results (from left to right)
(0–13,000 Mscf/year is equivalent to 0–368,160 m^3/year).

In order to identify the key parameters that affect CO$_2$ breakthrough, an analysis was performed to identify the key performance indicators (KPIs). According to this analysis, injector/producer pattern, relative location of the clusters, CO$_2$ breakthrough time indicator (calculated using the CO$_2$ data-driven smart proxy models for CO$_2$ breakthrough time prediction), and bottom-hole injection pressure are the essential parameters that control the amount of CO$_2$ production from the offset production well. Other important parameters that have a significant impact on CO$_2$ breakthrough include Langmuir isotherms for both CO$_2$ and CH$_4$, natural fracture permeability, and reserve-related parameters such as matrix and fracture porosity and pay thickness.

This emphasizes the fact that designed parameters such as well trajectory, hydraulic fracture placement, and operational constraints are among the parameters that should be optimized before starting the CO$_2$-EGR process.

A representative database with 116,000 input–output pairs was used to train a multilayer feed-forward back-propagation neural network. The data was partitioned with 70% training and 30% blind data (20% calibration and 10% validation). The designed neural network had one hidden layer with 52 hidden neurons.

The results of the development of the data-driven smart proxy for CO$_2$ production (Mscf/year) rates are shown in Figure 6.11. The cross plots in this figure show the predicted (by the smart proxy) and simulated values of CO$_2$ production rate for training, calibration, and verification steps (from left to right). In these plots, the x axis corresponds to the neural network predicted CO$_2$ production rate (Mscf/year), and the y axis shows the CO$_2$ production rates generated by the numerical reservoir simulator (Eclipse).

The data-driven CO$_2$-EGR&S smart proxy model was successfully developed to predict the CO$_2$ production rate (Mscf/year) with an R^2 of more than 0.99 for training, calibration, and verification. Figures 6.12 and 6.13 show examples of a comparison of the reservoir simulation output (blue dots) for CO$_2$ production rate (Mscf/year) during the CO$_2$-EGR&S process with the one predicted by the data-driven smart proxy model (red solid line) for some of the clusters and some of the laterals for all four production and injection patterns/cases. The plots clearly show that the developed data-driven

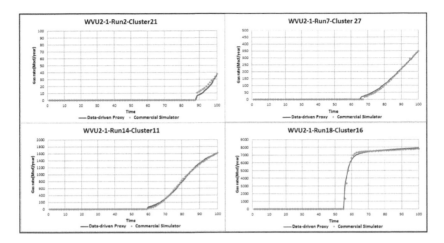

FIGURE 6.12
Comparison of CO$_2$ production rates (Mscf/year) from the simulator and CO$_2$-EGR&S data-driven smart proxy model for some of the hydraulic fracture clusters (Cases 1, 2, 3 and 4) (0–9,000 Mscf/year is equivalent to 0–254,880 m^3/year).

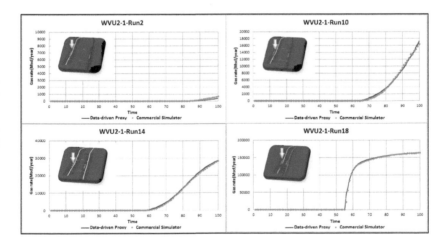

FIGURE 6.13
Comparison of CO$_2$ production rates (Mscf/year) from the simulator and CO$_2$-EGR&S data-driven smart proxy model for some of the laterals (Cases 1, 2, 3 and 4) (0–200,000 Mscf/year is equivalent to 0–5.66E+06 m^3/year).

CO$_2$-EGR&S smart proxy model is capable of accurately replicating the CO$_2$ production rate.

6.6.4 Blind Validation of the Smart Proxy Model

Blind validation of the developed data-driven smart proxy model is the final phase of the workflow to examine the capabilities of the smart proxy model

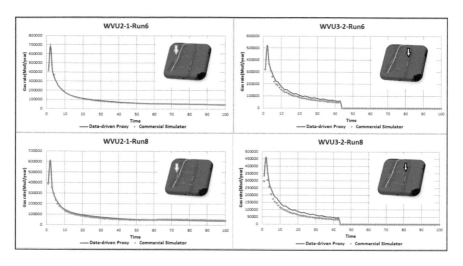

FIGURE 6.14
Comparison of CH$_4$ production rates (Mscf/year) from the simulator and the CO$_2$-EGR&S data-driven smart proxy model for some of the laterals: blind case (0–800,000 Mscf/year is equivalent to 0–2.55E+07 m^3/year).

in predictive mode. In order to accomplish this task, one of the cases (Case 2), including five simulation runs, was completely removed from the process of developing the smart proxy model (training, calibration, and validation of the models through 70%/30% partitioning of the spatio-temporal database). The same input parameters that were used for developing the four data-driven CO$_2$-EGR smart proxy models (CH$_4$ production, CO$_2$ injection rates, CO$_2$ breakthrough time, and CO$_2$ production rate prediction) were used to develop a new smart proxy model based on 15 simulation runs (instead of the initial 20 simulation runs).

On completion of the training, calibration, and validation of the new smart proxy models, the inputs (i.e., static properties and operational constraints) for the five simulation runs (with 58 clusters of hydraulic fracture per run) were introduced to the smart proxy models. The smart proxy model was deployed in forecast mode to generate the desired output for the blind simulation runs.

The corresponding outputs – including CH$_4$ production, CO$_2$ injection rates (Mscf/year), CO$_2$ breakthrough time (day), and CO$_2$ production rate (Mscf/year) – were generated, and some of the results are illustrated in Figure 6.14 (CH$_4$ production rate), Figure 6.15 (CO$_2$ injection rate), and Figure 6.16 (CO$_2$ production rate). In all the plots, blue dots are the outputs from the commercial numerical reservoir simulation model, and red solid lines are the results generated by the smart proxy model. As can be seen in Figure 6.14, the data-driven CO$_2$-EGR&S smart proxy model predicted the methane production rates for both production and the prospective injector with acceptable accuracy.

Since the hydraulic fracture length and number of clusters for WVU2-1 does not change for all 15 runs, the neural network could capture the

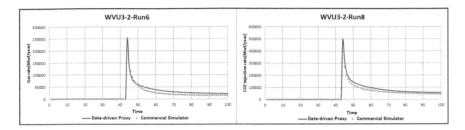

FIGURE 6.15
Comparison of CO_2 injection rates (Mscf/year) from the simulator and the CO_2-EGR&S data-driven smart proxy model for some of the laterals: blind case (0–800,000 Mscf/year is equivalent to 0–1.7E+07 m³/year).

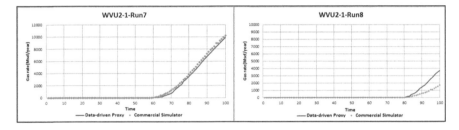

FIGURE 6.16
Comparison of CO_2 production rate (Mscf/year) from the simulator and the CO_2-EGR&S data-driven smart proxy model for some of the laterals: blind case (0–12,000 Mscf/year is equivalent to 0–339,840 m³/year).

production behavior of this well better than WVU3-2. Figures 6.15 and 6.16 show a comparison of the CO_2 injection and CO_2 production rates from the offset producer calculated by the commercial numerical reservoir simulation model with the profiles predicted by the smart proxy model for some of the runs, which were completely blind to the neural network. According to these figures, the data-driven smart proxy model was able to mimic the simulation behavior with good accuracy.

The second method used to validate/test the predictive capabilities of the developed data-driven smart proxy model was to design a completely new horizontal well with unique reservoir and hydraulic fracture characteristics for each cluster and with new operational constraints that were completely different (but in the uncertainty range) from the initially designed 20 simulation runs used for the training, calibration, and validation of the neural network. Therefore, a new synthetic well (WVU-2013) was drilled and completed with 25 clusters of hydraulic fractures.

Figure 6.17 shows the new horizontal well trajectory in orange and its position relative to the other offset laterals in the WVU pad. The simulation outputs were compared with the data-driven smart proxy model results. The results of these comparisons are shown in Figure 6.18. In this figure,

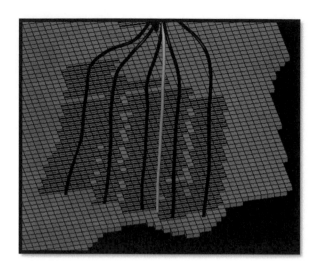

FIGURE 6.17
Schematic of new synthetic well for further validation of the smart proxy models.

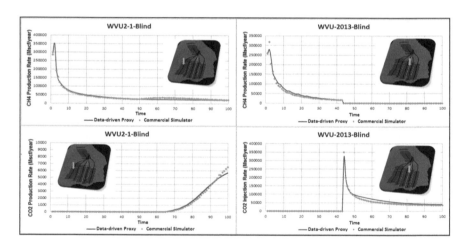

FIGURE 6.18
Comparison of CH$_4$, CO$_2$ production profiles (Mscf/year) and CO$_2$ injection rates (Mscf/year) from the simulator and the CO$_2$-EGR&S data-driven smart proxy model for some of the laterals: blind case (0–400,000 Mscf/year is equivalent to 0–1.133E+07 m^3/year).

the top plots compare the simulation and smart proxy model results for CH$_4$ production for the producing and prospective injector wells. The bottom left plot shows CO$_2$ production and the bottom right figure shows the CO$_2$ injection profile comparisons.

These blind validation results clearly demonstrate that the developed data-driven CO$_2$-EGR&S smart proxy models can be used as an efficient and fast uncertainty quantification tool, because thousands of proxy model runs can be made quickly (in minutes) to fulfill any uncertainty analysis, techno-economic evaluation, and optimization purpose.

6.7 Summary and Conclusions

This chapter has introduced shale as a potential geological storage for CO_2. Difficulties associates with the numerical simulation of CO_2 injection into Marcellus shale are discussed, and a practical solution to address those problems is proposed. A series of data-driven CO_2-EGR&S smart proxy models for shale formation are developed based on the pattern recognition capabilities of artificial intelligence and validated by completely blind simulation runs to reproduce the injection and production profiles for the CO_2-EGR&S process.

This technique provides the ability to perform fast, detailed uncertainty and optimization analysis, instead of using a numerical simulator for which the model setup and implementation are laborious and computationally expensive. Additionally, analyses of the KPIs are performed using the fuzzy pattern recognition technique, and the most influential parameters that control CH_4 production rates, CO_2 injection rates, CO_2 breakthrough times, and CO_2 production rates are identified as important considerations before designing any CO_2 injection process for the purpose of EGR&S.

7

CO_2-EOR as a Storage Mechanism

Shahab D. Mohaghegh, Alireza Shahkarami, and Vida Gholami

CONTENTS

It is an established fact that injection of CO_2 into oil reservoirs has the potential to improve oil recovery. CO_2 behaves as a supercritical fluid above its critical temperature (31.10°C, 87.98°F) and critical pressure (7.39 MPa, 1,071 psi) as shown in Figure 7.1. Under most reservoir conditions, CO_2 is in a supercritical state. The supercritical conditions of CO_2 make it a great choice as an agent for a process that in the oil industry is referred to as enhanced oil recovery (EOR). In its supercritical state, CO_2 has high solvency power to extract hydrocarbon components for the miscible displacement of oil. The miscible state of multiple fluids refers to their homogenous mixture.

Performing the CO_2-EOR process in depleted oil reservoirs is both economically and environmentally beneficial, because from an oil recovery point of view, upon completion of the primary recovery of an oil reservoir (which may be referred to as depleted), large amounts of oil (more than already produced) are left behind. Therefore, CO_2-EOR can work as an immediate option to reduce CO_2 emissions into the atmosphere in combination with enhancing oil recovery. The aim in the CO_2-EOR process is to minimize the CO_2 required to produce a barrel of oil, while for sequestration purposes,

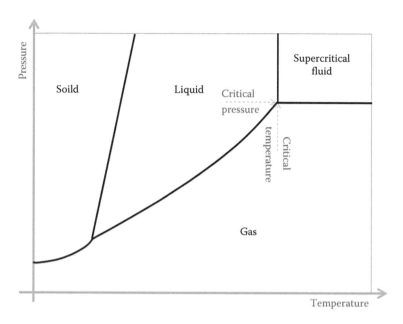

FIGURE 7.1
Different phases of matter. Pressure and temperature control the state of a matter from solid, to liquid, to gas, and to super critical state.

maximizing the amount of CO_2 stored is the objective (70). This dichotomy calls for an optimization process such that the maximum amount of CO_2 is stored in the reservoir while incremental oil recovery is also maximized.

CO_2-EOR was initially tried in 1972 in Scurry County, Texas. Since then, this approach has been used successfully throughout the Permian Basin of West Texas and eastern New Mexico, and is now being pursued to a limited extent in Kansas, Mississippi, Wyoming, Oklahoma, Colorado, Utah, Montana, Alaska, and Pennsylvania.

Many factors, including wettability, gravitational effects, and reservoir heterogeneities, affect the premature breakthrough of CO_2, which in turn strongly influences the recovery of the oil it is trying to displace. Furthermore, operations onshore and offshore have their specific characteristics, offshore being more complicated, with higher costs and the requirement for more advanced technologies (71). If CO_2 is injected at a pressure below the minimum miscibility pressure (MMP), then it can enhance pumping by swelling the oil and reducing its viscosity. MMP is the lowest pressure at which miscibility between two or multiple fluids occurs. At the MMP, the interfacial tension is zero and no interface exists between the fluids.

There are several reasons why CO_2-EOR is a favorable method of sequestration. First, given the integrity of the geological structures that originally held the oil and gas, they should also permanently contain the injected CO_2. Since operating depleted oil and gas fields are widespread,

TABLE 7.1

CO$_2$ Utilization and Potential in EOR Projects

USA (2006)	Amount
CO$_2$ use for EOR (millions of tons/year)	48
• Naturally occuring	36
• Anthropogenic	12
Estimated CO$_2$ sequestered by EOR operations	9
Worldwide	
Potential CO$_2$ EOR sequestration	130 billion tons
Total CO$_2$ accumulated in the atmosphere	3–4 billion tons/year

Source: U.S. Department of Energy. 2007. *U.S. Department of Energy Carbon Sequestration ATLAS of the United States and Canada.* Morgantown, WV: US DOE National Energy Technology Laboratory.

there is a good likelihood of them being close to a CO$_2$ source. Finally, carbon sequestration from CO$_2$-EOR projects can generate counterbalances occasioning trades in the evolving greenhouse gas market (Table 7.1) (72). According to the U.S. Department of Energy (DOE), depleted oil and gas wells in the United States and Canada can potentially sequester over 82 billion tons of carbon in total (73).

In this chapter we present a numerical reservoir simulation and modeling of a comprehensive CO$_2$-EOR project, including the application, utility, and important contributions of data-driven analytics in projects involving CO$_2$-EOR. The chapter includes a geologic and operational description of the CO$_2$-EOR project in West Texas, followed by the development and analyses of the numerical simulation and modeling for the project and then the incorporation of smart proxy modeling, which maximizes the utility of the numerical modeling study.

7.1 The Field and Some Background

The Kelly-Snyder field (Figure 7.2), discovered in 1948, is one of the major oil reservoirs in the United States. The original estimate of its original oil in place (OOIP) was approximately 2.73 billion bbls. The primary production mechanism was indicated as merely solution gas driven, based on the early performance history of the field, which would probably result in an ultimate recovery of less than 20% of the OOIP. The Scurry Area Canyon Reef Operations Committee Unit (SACROC Unit) was formed in 1953, and in September 1954 a massive pressure maintenance program was started. Based on this plan, water was injected into a center-line row of wells along the longitudinal axis of the reservoir (74).

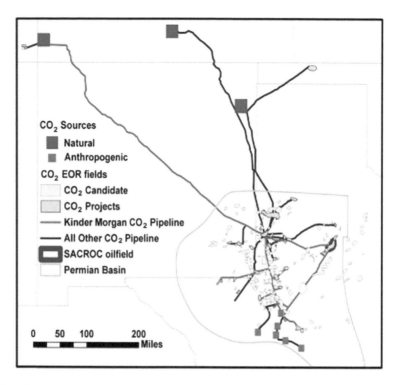

FIGURE 7.2
The Permian Basin outlined in green covers parts of western Texas and southeastern New Mexico. SACROC oilfield, identified in red, resides on the northeastern edge of the Basin. The map also identifies both natural and anthropogenic sources of CO_2 as well as the CO_2 pipelines in the region. (Bureau of Economic Geology, The University of Texas at Austin, Gulf Coast Carbon Center, 2017, http://www.beg.utexas.edu/gccc/sacroc.php)

In 1968, a technical committee, studying potential substitutes, suggested using a water-driven slug of CO_2 for the miscible displacement of the oil in the non-water-invaded portion of the reservoir. It was also recommended that a pattern injection program be developed in this area, to apply the slug process and enhance ultimate oil recovery. CO_2 injection began in early 1972. An inverted nine-spot miscible flood program, consisting of injecting CO_2 driven by water, was decided to be the most effective and economical alternate method to improve recovery in the SACROC Unit based on the investigations. A comparison of the predicted ultimate recovery of the original water injection program and the recommended scheme showed that the new pattern would result in an increase of about 230 million bbls in production and recovery (74).

The SACROC Unit, within the Horseshoe Atoll, has undergone CO_2 injection since 1972 and is the oldest continuously operated CO_2-EOR operation in the United States. The amount of injected CO_2 until 2005 was about 93 million tons (93,673,236,443 kg), of which about 38 million tons (38,040,501,080 kg) has been produced along with the recovered oil. Accordingly, a simple mass balance suggests that about 55 million tons (55,632,735,360 kg) of CO_2 have been accumulated in the site (75). Initially, Chevron Corporation operated the unit. Later, Pennzoil took over before its upstream operations were spun off into Pennz Energy Co. and merged with Devon Energy Corporation. In June 2000, Kinder Morgan CO_2 bought Devon's interest in the SACROC Unit. At the time of Kinder Morgan's purchase, the production, which once was more than 200,000 bbls/day, had declined to 8,200 bbls/day. Kinder Morgan continues CO_2 flooding of the SACROC Unit.

Since the SACROC project was the first large-scale CO_2-EOR, a lot of lessons were learned during its operational history. Chevron did not have access to sufficient CO_2 to perform flooding throughout the entire field during a reasonable time frame. The parts of the reservoir that performed poorly during the water flooding recovery and had the highest percent of OOIP were prioritized. It took years of experience in multiple projects to reach an industry consensus that the areas that have performed the best under the water flooding process will also have the best response to CO_2 flooding. Variation in well-to-well connectivity was found to be the main reason for the difference in performance. The higher amount of remaining oil in the formation in some areas was of little use if the injected fluid never contacted it (76).

7.2 The Reservoir

Of the four contiguous fields along the 35×5 mile (56.3×8.1 km) Canyon Reef Formation, the Kelly-Snyder field, in Scurry County (West Texas) is the most important unitized field. The formation is a Pennsylvanian age limestone occurring at an average depth of 6,700 ft (approximately 2,040 m). It has a northeast–southwest trending massive reef build up with thinner, gently sloping flanks. The formation thickness varies from an average of 800 ft (244 m) on the crest of the structure to less than 50 ft (15 m) on the flanks, and averages 213 ft (65 m) overall.

The SACROC Unit covers just less than 50,000 acres and constitutes about 98% of the Kelly-Snyder field. The unitized reservoir is generally bounded on the east and west by porosity development and on the northeast and southwest by offsetting units. An oil–water contact, although poorly defined in some areas, occurs throughout most of the Unit area. Only very limited water influx is detected during pressure depletion, indicating a relatively small aquifer in the immediate area (74).

Reservoir oil was initially under-saturated at an original pressure of 3,122 psig (21,625 kPa), and had a solution gas content of slightly less than 1,000 scf/ STB (stock tank barrels) and a bubble point pressure of 1,805 psig (12,546 kPa). The reservoir oil is rich in intermediate components (31.5 mole percent C_2–C_4)

TABLE 7.2

Basic Reservoir Data – SACROC

Parameter	Amount
Approximate depth (m)	2,042
Approximate water oil contact (WOC) (m) subsea	−1,370
Average gross thickness (m)	65
Average porosity (%)	7.6
Average permeability (mD)	19.4
Average interstitial water saturation (%)	36
Average residual oil saturation (%)	26
Reservoir temperature (°C)	55
Initial reservoir pressure (kPa)	21,627
Water injection commencement	September 1959
Total area (acres)	49,900
Total wells in the unit	1,256
Unit hydrocarbon pore volume (MMbbls)	4,014.253
Original oil in place (MMbbls)	2,727

Source: Dicharry, R. M., Perryman, T. L. and Ronquille, J. D. R. 1973. Evaluation and Design of a CO2 Miscible Flood Project – SACROC Unit, Kelly Snider Field. *Journal of Petroleum Technology*, SPE4083.

(74). While the OOIP within the SACROC Unit area, as mentioned before, was estimated to be 2.73 billion STB, cumulative production through 1971 was 536 million STB, or approximately 19.7% of the OOIP. Basic reservoir data for the SACROC Unit is shown in Table 7.2 and the fluid properties are shown in Table 7.3.

7.3 Geologic Description

SACROC is located in the southeastern segment of the Horseshoe Atoll within the Midland basin of West Texas (Figure 7.3). SACROC Unit covers an area of 356 km^2 with a length of 40 km and width of 3–15 km, all within the Horseshoe Atoll (77). Geologically, massive amounts of bedded bioclastic limestone and thin shale beds representing the Strawn, Canyon, and Cisco formations of the Pennsylvanian, and the Wolfcamp Series of the Lower Permian, comprise the carbonate reef complex at SACROC (77). Among these formations, most of the CO$_2$ for EOR is injected into Cisco and Canyon formations, which were deposited during the Pennsylvanian age.

The Strawn Formation in the carbonate reef complex started to form in the early Desmoinesian period, while the basin was on the equator (79). The carbonate sedimentation of the Canyon and Cisco formations continued

TABLE 7.3

SACROC Reservoir Fluid Properties

Fluid Parameter	Amount
Bubble point pressure @ 55°C (kPa)	12,604
Fluid viscosity @ 55°C (cP)	0
Fluid density @ 55°C (g/cc)	0.67

Flash separation data – First stage separator conditions

	172.4 kPa and 35°C	213 kPa and 23.9°C
Solution gas oil ratio (scf/STB)	990	910
Stock tank oil gravity (°API)	41	42.7
Casing head gas gravity	1.087	1.03
Formation volume factor @ 21,629 kPa (bbls/STB)	1.53	1.42
Formation volume factor @ 12,548 kPa (bbls/STB)	1.56	1.5

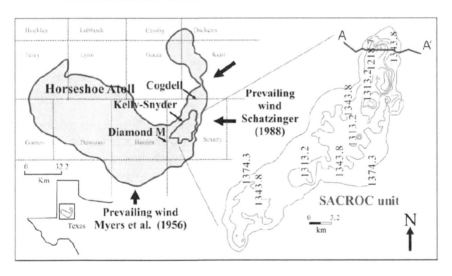

FIGURE 7.3
SACROC Unit at the Horseshoe Atoll in West Texas and structural contours map of the top of the carbonate reef modified from Stafford. Contours are in metres. (From Myers, D.A., Stafford, P.T., Burnside, R.J., 1956. Geology of the late Paleozoic Horseshoe Atoll in west Texas. Bureau of Economic Geology Publication 5607, p.113 (307).)

during the Missourian and Virgilian periods. Drastic influx of fine-grained clastics ended the accumulation of carbonate sediments on the SACROC during the Wolfcampian period. Although the Canyon and Cisco formations are mostly composed of limestone, minor amounts of anhydrite, sand, chert, and shale are present locally (75). Based on an analysis done by Carey et al.

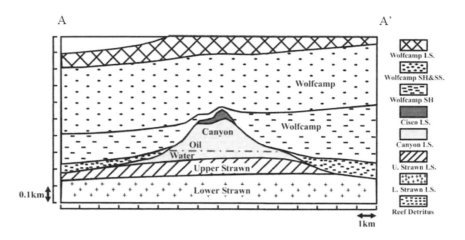

FIGURE 7.4

A structural and stratigraphic cross-section of profile A-A′, located within the SACROC northern platform. See Figure 7.3 for location of profile A-A′., AAPG ©[2017], (reprinted by permission of the AAPG whose permission is required for further use.)

on core samples from wells 49–5 and 49–6 in the SACROC field, the limestone is mostly calcite with minor ankerite, quartz, and thin clay lenses (80).

The Wolfcamp shale of the lower Permian acts as a seal above the Canyon and Cisco formations (75). According to X-ray diffraction analysis results, the shale is mostly illite/smectite and quartz with minor feldspar, carbonate, and pyrite. Carey et al. concluded that the CO_2 had not interacted with the shale, based on mineralogical analysis (80).

A lot of investigations have been performed by petroleum geologists to understand the spatial (areal) distribution of carbonates in the SACROC Unit, along with the variation of carbonates with depth. Based on these analyses, this unit has been divided into Lower Canyon, Middle Canyon, Upper Canyon, and Cisco groups through detailed analyses of cores, logging data, and biostratigraphy. These investigations suggested that the Cisco group unconformably overlies the Canyon Group (81). A structural and stratigraphic cross section of the SACROC is shown in Figure 7.4. Both the porosity and permeability in SACROC Unit show a large variability due to changes in the depositional environment. The values of SACROC porosity and permeability reported in previous studies are provided in Figure 7.5.

7.4 Performance History

The Standard Oil Company of Texas discovered the SACROC Unit in November 1948. The OOIP in this unit was estimated to be approximately 2.73 billion STB in the Canyon Reef limestone formation (74). The rapid

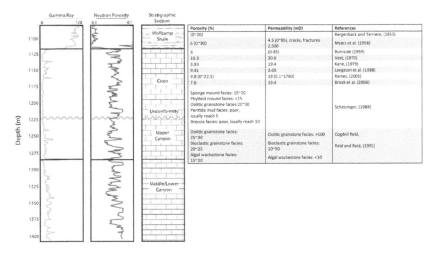

FIGURE 7.5
Summary of previous studies estimating carbonate rock properties in Cisco and Canyon Groups. (Han W.Sh. Evaluation Of CO$_2$ Trapping Mechanisms At The Sacroc Northern Platform: Site Of 35 Years Of CO$_2$ Injection [Report]: PhD Dissertation – Socorro, New Mexico: The New Mexico Institute of Mining and Technology, 2008.)

development of the field is obvious from the oil production rate buildup during the first 2 years. Over 1,200 producing wells with 81 operators were drilled in the Canyon Reef complex from 1948 to 1951 (82).

A study of this early performance showed that the solution gas drive was the primary reservoir mechanism and no effective water drive existed. During the first five years, less than 5% of the OOIP was produced; however, the gas oil ratio (GOR) increased and the average reservoir pressure was reduced by 50% (i.e., from 21,525 to 10,755 kPa). It became obvious that some form of pressure maintenance was needed to prevent very low oil recovery. Relying on only a solution gas drive would result in an ultimate recovery of only 19% of OOIP, based on performance predictions. The field was unitized in March 1953 and has since been operated as the SACROC Unit.

7.5 Water Flooding Evaluation

In September 1954, a full-scale program of pressure maintenance by water injection throughout the unit area began in SACROC. The first 72 water injection wells, with an injection rate of 132,000 barrels of water per day (BWPD), were selected generally along the longitudinal crest of the structure. The wells in the thicker part of the reservoir were selected bearing in mind the critical pressure and gas saturation conditions that existed and the obvious necessity for the speedy restoration of pressure describing a "center-line pattern" injection (74).

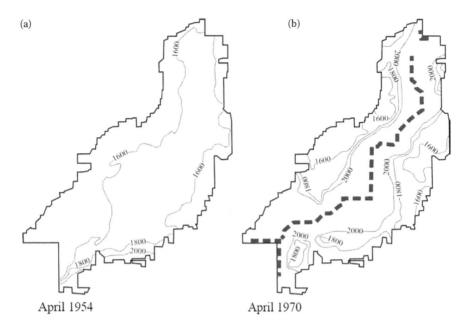

April 1954   April 1970

FIGURE 7.6

Contour and dotted lines, respectively, indicate the bottom-hole pressure (psi) and the location of center-line water injection wells. (a) Five months before starting water injection only 1% of reservoir volume is above bubble point pressure (1,805 psig = 12,445 kPa). (b) Seventeen years after starting water injection, 77% of the reservoir volume is above bubble point pressure.

As shown in Figure 7.6, before water injection began, only 1% of the reservoir was above bubble point pressure (1,805 psi/12,445 kPa). Within less than two years, the pressure in 45% of the reservoir rose above bubble point pressure. Finally, 77% was above bubble point pressure after 17 years of water injection (74).

7.6 Enhanced Oil Recovery Methods

EOR refers to processes that are usually incorporated once primary recovery is close to its end. During primary recovery the oil reservoirs are produced using the natural, initial energy (pressure) stored in the reservoir. During the EOR process, new energy is introduced into the reservoir in multiple forms to ease and assist the flow of the remaining oil in place.

7.6.1 Enriched Miscible Gas Process

The enriched miscible gas process did not look attractive for the SACROC Unit due to the high cost of the injected material and the risks involved with gravity override and viscous fingering. The CO$_2$ slug process was therefore selected as the most appropriate EOR process in the SACROC Unit.

7.6.2 Carbon Dioxide Process

Laboratory work and field studies (83,84) have demonstrated that, under favorable conditions, CO$_2$ miscible displacement is a good and productive EOR method. Laboratory research also showed that CO$_2$ would successfully generate a miscible displacement of the SACROC oil under bubble point conditions. Successive laboratory work indicated that the injected CO$_2$ gas could be diluted with as much as 18% methane without seriously affecting the miscibility-generating capabilities of the injected gas mixture.

This was a key point, because CO$_2$ in the volumes required was not available in a pure form, but only as a methane-diluted waste by-product from CO$_2$ removal plants designed to produce marketable hydrocarbon residue gas. The injection pattern of the inverted nine spots was designed to process approximately 49% of the original unit hydrocarbon pore volume. It was decided that the EOR process be performed based on different phases because of supply rate limitations.

7.7 Development of Numerical Reservoir Simulation

There are various questions that can be addressed using numerical simulations, including but not limited to the subsurface volume to be affected by the injected CO$_2$, the effect of formation heterogeneity, gravity override, and viscous instability on the utilization of subsurface space, the fraction of CO$_2$ trapped in the subsurface in different states and the mechanism of trapping, evolution of leak and what causes it, the role of different rock and fluid properties such as relative permeability and capillary pressure on CO$_2$ containment and leak, the role of different phase compositions and phase changes in CO$_2$ leakage (85).

Although many of the simulation techniques for multi-phase flow, commonly used in the industry, are applicable to CO$_2$ storage and sequestration, there are new subtle problems that should be addressed, such as the great scale of time and space affecting CO$_2$ behavior in subsurface geologic sites, the integration of fluid flow, chemical and geomechanical processes, the complexities of near critical behavior, which become important while CO$_2$ leaks, and integrating interfacial processes such as lithological contacts and ground surface (86).

It is important to understand the reaction between the water–gas and minerals in a CO$_2$ storage and sequestration process. As there are many limitations to laboratory or field tests, one of the supplementary or independent tools widely used to accomplish this is reactive transport modeling. The limitations arise from the complexity involved with coupled hydrodynamic, chemical, mechanical and thermal processes during CO$_2$ injection or leakage from a geological site. NUFT (87), FLOWTRAN (88,89), GEM (90), and TOUGHREACT (91) are among the computational reactive models that have been specifically developed for numerical simulation of carbon storage and sequestration (85).

The numerical reservoir simulation model for the SACROC Unit that is presented in this chapter was developed based on the original work by a previous researcher (92). The original reservoir model was for a CO$_2$ EOR project lasting for 200 years, from 1972 to 2172. The model utilized in this study covers the period from January 1, 2172 to January 1, 3172, after the reservoir has been depleted of oil. Therefore the simulation model is just considered for CO$_2$ storage and sequestration. The model contains 25 simulation layers of 16 × 34 grid blocks. There are 45 injection wells planned, to inject CO$_2$ at a constant rate (331,801.9 m^3/day) for 50 years starting in 2172.

Each well is perforated in a single layer, although the perforated layers might be different for different wells. The perforations happen in layers 19 (one well), 20 (40 wells), 21 (one well) and 22 (three wells). It is assumed that there are no-flow boundary conditions at the outer boundaries. Figure 7.7 shows a three-dimensional view of the structure of the reservoir modeled using the numerical simulation method.

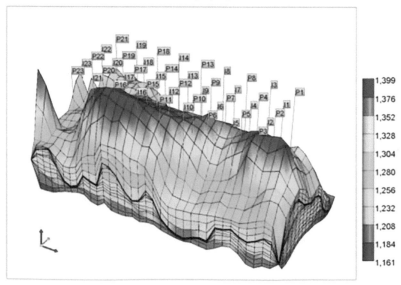

FIGURE 7.7
A 3-D view of the simulation model used for the study in this chapter.

The objective of this reservoir model is to track the distribution of the pressure and phase saturations at the target layer (layer 18) during and after the injection of CO$_2$. The total number of grid blocks in this layer is 544, of which only 422 are active. The initial properties (pressure, water saturation, and gas mole fraction [CO$_2$], respectively) at layer 18 are shown in Figures 7.8 through 7.10. The white grid blocks are "null" or inactive because they have a negligible thickness value. The initial condition is the condition of the reservoir after 200 years of the EOR process (from 1972 to 2172), which comes from the original model.

This numerical reservoir simulation model is the subject of the smart proxy model that is the actual focus of this chapter as an implementation of data-driven analytics in CO$_2$ storage problems. The remainder of this chapter is dedicated to details of the development and then the results associated with the development and deployment of smart proxy modeling for CO$_2$ storage in geological formations.

FIGURE 7.8
Initial pressure at the target layer (layer 18) for the base simulation model.

FIGURE 7.9
Initial water saturation at the target layer (layer 18) for the base simulation model.

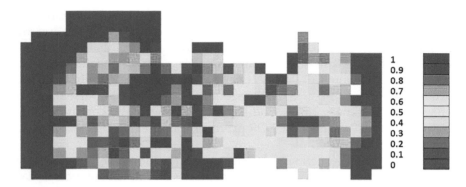

FIGURE 7.10
Initial gas (CO$_2$) mole fraction at the target layer (layer 18) for the base simulation model.

7.8 Proxies of the Numerical Simulation Models

It is not uncommon to carry out hundreds of numerical simulation runs for model updating, uncertainty analysis, or optimization steps. In examples that address non-academic, real-world problems, a single simulation run can take tens of hours or even days, and parallel processing on tens of CPUs may be incorporated. Recently, proxy models have been used quite extensively to alleviate the computational cost associated with many of the problems in the petroleum industry, including field development, history matching, uncertainty analysis, optimization, and so on.

While some methods have been used for well-based proxy modeling, there are very few techniques available in the literature that focus on grid-based proxy modeling for a black-oil reservoir model, let alone compositional simulations. These methods have severe pitfalls and drawbacks. Attempts have been made to shed some light on these methods and are described in the following paragraphs.

Lerlertpakdee et al. (93) used the relationship between the output performance of the reservoir with the input parameters and the production strategy to develop a reduced-order model (ROM) in a two-phase flow system. Their model was a one-dimensional (1-D) flow network, which essentially used the injection/production information. In this method, instead of considering all the grid blocks and solving the partial differential equations (PDEs) only the connections of the wells are considered and divided into course blocks and the derived system of equation is solved for the new 1-D flow network model.

The calibration process in this study includes calculating the output of multiple high-fidelity model runs followed by using an objective function, which minimizes the misfit between the high- and reduced-order model results by changing the width (representing pore volume, PV) and permeability of each grid block. Although the model has shown a close match between the high-order and reduced-order model of net present value (NPV), it should be noted that the only variables that are changing in the 10 training sets are the bottom-hole pressures (BHPs) of the wells in a specific range.

The models used for this study were small (maximum 13,200 grid blocks) and two-dimensional (2-D). A potential computational cost increase or accuracy loss due to gross upscaling might result if the technique is used for multiphase flow and a more complex geological model with multiple layers and production/injection wells.

Zhang and Sahinidis (94,95) used polynomial chaos expansion (PCE) to build a proxy model, which is used for uncertainty quantification and injection optimization in a carbon sequestration system. The defined model is a 2-D homogenous and isotopic saline formation, with 986 grid blocks. Quantifying the impact of the uncertain parameters (porosity and permeability) on the model outputs (pressure and gas saturation throughout the formation) were of interest. One hundred simulation runs were performed in order to build PCEs. The PCEs are of order $d = 3$ or 4.

Since the coefficients in the PCE model are dependent on space and time, this process involves generating a large number of PCE models; that is, for each output (e.g., pressure) the number of required PCEs will be the product of the number of grid blocks and the number of time steps. As the number of inputs (uncertain parameters) increases, the number of required numerical runs grows sharply, which makes this approach impractical; for example, at least 8,008 runs are required in the case of having eight inputs and PCE of order 6.

Runtimes of seconds for PCE and 15 minutes for the numerical simulation have been reported, and it should be noted that each PCE provides information only in one grid block and for a specific time, whilst the numerical simulation provides the results at all time steps and all grid blocks. In gas saturation calculations, some of the outputs are negative values, which are numerical errors due to approximations using the polynomial terms. For gas saturation, after 30 days of injection, an average error of 6% has been reported. Although using contours has smoothed the results, there is some wiggling on the pressure contour boundaries, which might be due to the oscillating feature of polynomial terms in the expansion.

Van Doren et al. (96) proposed reduced-order modeling for production optimization in the water flooding process by using proper orthogonal decomposition (POD). They constructed reduced-order models (ROMs) using the data obtained from many snapshots of the model states (pressure and water saturation distribution) from the full-order simulation model. The POD was used to summarize the dynamic variability of the full-order reservoir model in a reduced subspace. Although the number of state vectors (containing oil pressures P_o and water saturations S_w for each grid block) is decreased using POD, the change in the matrix structure from penta-diagonal (or hepta-diagonal in 3-D systems) to a full matrix counteracts the computational advantage obtained by vector size reduction.

The methodology was tested on a water flooding scenario in a 2-D, two-phase model with 2,025 grid blocks and two horizontal (injection and production) wells with control valves in each grid block (90 segments). The resulting reduction in computing time for NPV optimization for this sample model was reported to be 35%, at best. When the ROM was simulated with the same controls (rates of wells) as the original full-order model, the states were almost identical. However, if the controls were altered (blind test), then the states of the full-order model were less represented by the ROM.

In general, although the POD methodology yields ROMs with low complexity, the actual speed up for the simulation is modest as compared with the size of the models. This is due to the fact that the nonlinear function for estimating the state vectors will still be evaluated at the full-order number of states, which can be computationally expensive and inefficient. The main drawbacks of the POD stem from the fact that the projection basis depends on the training inputs and the timescale at which the snapshots are taken.

In 2009, Cardoso and Durlofsky (97) emphasized that, for nonlinear problems, the POD procedure is limited in terms of the speed up it can achieve, because it targets only the linear solver, and the computation effort for some of the operations (constructing the full residual and Jacobian matrices at every iteration of every time step) is not reduced. They applied a linearization process (trajectory piecewise linearization, TPWL) to the governing equations in addition to the ROM obtained from the POD projection, and incorporated it in production optimization.

The process under study was water flooding in a 3-D model having 20,400 grid blocks and six wells (four producers and two injectors), in which the well BHPs were altering. The TPWL algorithm was first developed in the electrical engineering framework in order to extract ROMs in a circuit simulation. Although the final calculations for estimating the new state vectors can be performed quickly using this method, the preprocessing calculations, including running some number of high-fidelity training simulations, the construction of the reduced basis function, construction and inversion of the reduced Jacobian matrix for all saved states (which needs modification in the simulator code), construction of a reduced representation of states, an accumulation matrix, and source/sink term matrix of derivatives, are still computationally costly.

On the other hand, the accuracy of the TPWL solution is sensitive to the number of basis vectors used in the projection matrix; therefore, some amount of numerical experimentation may be required to establish these numbers for saturation and pressure. Instabilities and a deterioration of accuracy due to application of the procedure to control inputs far away from the training trajectories are two of the main drawbacks of TPWL (98).

One application of proxy models is to deal with the compositional simulation. These simulations are required for modeling the EOR, CO$_2$ sequestration, or gas injection. Due to the intrinsic nonlinearity of these models and the potentially large system of unknowns, they can demand high computational powers. For production optimization, in particular, hundreds or thousands of simulation runs should be performed, which accentuates how imperative it is that a model runs efficiently.

Proxy models sometimes have been used as a term just to represent correlations or simple mathematical models to provide single outputs. In one of these applications, the oil recovery of a reservoir after CO_2 flooding has been modeled as a hint for reservoir screening. The reservoir models used in this study are simple homogenous models (except in permeability), which differ in the type of reservoir fluid (99).

In 2009, Yang et al. (100) used a hybrid modeling technique for reservoir development using both full-physics and proxy simulations. The proxy model used in their work is a profile generator. It simplifies the reservoir into material balance tanks, and well source/sink terms into a set of tables known as type curves, which relate the production GOR, water cut, etc., to the estimated ultimate recovery (EUR) and other parameters. Their objective is to reduce the computational cost associated with the modeling of huge reservoirs. The profile generation data should be provided for each well, which is not using the simulation grid. The main assumption is that the reservoir is to be operated in a way not significantly different from the base case for which the type curve data is produced, which might not be practical in most real field problems.

In another attempt at using proxy models, NRAP (National Risk Assessment Partnership) focused on using this tool for long-term quantitative risk assessment of carbon storage (101). This is performed by dividing the carbon storage system into components (reservoir, wells, seals, groundwater, and atmosphere), using proxy models for each component, and integrating all the models to assess the success probability of carbon storage using the Monte Carlo simulation. Different proxy models are used, including Look-Up Table (LUT), response surface, PCE, and artificial intelligence (AI)-based surrogate reservoir models.

The look-up table methodology is very simple but requires hundreds of runs of the high-fidelity model based on different inputs. The table is built based on the inputs, the results of the simulation runs, and a third dimension representing the time step. The outcome of a new scenario can be obtained using an interpolation-based approach from the created table. Although this method is quite rapid, the problem lies in the number of full-physics model simulations needed to build the table. In the work performed by NRAP, for a 2-D, two-phase (saline formation) model with 10,000 (100 × 100) grid blocks, and only three variable parameters of reservoir permeability, reservoir porosity, and seal permeability, more than 300 runs were needed to build a table to predict the pressure and saturation at each grid block.

A heterogeneous field cannot be used in this approach, and permeability must be varied through a scalar multiplier. Different snapshots of time were selected in the interval of 1,000 years of post-injection. The size of the look-up table and accuracy of the model depends on the selected snapshots and the time span between them.

In 2014 He and Durlofsky (102) used POD-TPWL, a combination of TPWL and POD, in order to build a ROM. Using POD alone will result in high-order complexity due to the need for the construction and projection of nonlinear terms; TPWL has been used to address this problem. One of the limitations with this method is that the system stability is highly dependent on the type of projection scheme used. This method needs a lot of offline processing to construct the POD-TPWL model.

In order to obtain more accurate results, more training runs are needed, which results in storage and computational problems. Modifications should be done to the reservoir simulator; hence it is not applicable using conventional reservoir simulators. The number of variables is equal to the number of grid blocks multiplied by the fluid components. Pre-processing (offline computations) involves running the full-order training simulations, saving and reading the states and derivatives, constructing the basis matrices, and reducing the states and derivatives, which for only two or three training simulations requires the same amount of time as one full-order simulation run.

This technique has been tested on a small synthetic reservoir with a few wells (less than 10). The primary variables were set to be pressure and the component mole fraction. To obtain the flow rate, the full-order primary variables should be reconstructed at a specific time and location, and secondary variables, such as saturations, should be calculated by performing flash calculations with the primary variables. Increasing the number of wells will introduce more variability in the states, which may have a considerable effect on the model results and computational expense. Besides, since the solution is made based on linearization around the generated states, it is extremely important that the test case is in the range of and close to the training runs. For instance, the variable parameter in the test cases (BHP) is selected to be very similar to the training cases.

In 2012, Zhang and Pau (94) developed a ROM for CO$_2$ storage in brine reservoirs. Their objective was to use the ROM for risk assessment of geological carbon sequestration. Their study was based on building a response surface from a set of high-fidelity forward simulations for selected parameter values. The approaches used included Gaussian process (GP) regression, radial basis function, and a look-up table combined with linear interpolation.

Their ultimate goal was to predict the pressure value at a specific location and specific time in the reservoir. A relative error was used to quantify the accuracy of the approximation. Only three parameters (permeability and porosity of the sand layer and permeability of the cap-rock) were used to build the realizations. While the highest number of simulation models used for this approach is 57, which is much less than a typical response surface approach, the key limitation of this method is that the ROM will be valid for predicting only one output of interest (e.g., for a specific time and location).

Thus, in order to predict each parameter throughout the reservoir over a span of time, the number of ROMs required to be built will be equal to the number of grid blocks multiplied by the number of time steps, which makes this approach prohibitive.

In another work that was performed in 2003, Hejin et al. (103) studied the application of five different methods in order to derive low-order models of two-phase (oil–water) reservoir flow. Their study was performed on a simple synthetic model with eight grid blocks in the X and Y directions and only one layer. Two wells (one injection and one production) were considered. Modal decomposition, balanced realization, a combination of the last two methods, and subspace identification resulted in linear low-order models, which were only valid for a limited time span during which the linearization was valid (only 10 days in their study). POD reduced the high-order model to a nonlinear low-order model; however, it did not reduce the simulation time.

In 2013, Gildin et al. (98) combined the discrete empirical interpolation method (DEIM) with POD and proposed using this method to overcome the issues with the nonlinear projections. DEIM is based on the approximation of the nonlinear terms by way of an interpolatory projection of a few selected snapshots of nonlinear terms. The authors applied this model on a 2-D, two-phase, and five-spot reservoir model with about 2,000 grid blocks and compared the results in terms of oil pressure and water saturation with POD-TPWL. POD-TPWL gave a faster methodology and smoother results when approximating the full nonlinear behavior of the reservoir in this case. Oscillations were obtained in the production profiles of oil and water for the five-spot pattern studied, which invites more study on the interpolation algorithm.

Data-driven modeling has sometimes been used in combination with other reduced-order modeling techniques. Klie (104) used a non-intrusive model reduction approach based on POD, DEIM, and radial basis function (RBF) networks to predict the production of oil and gas reservoirs. POD and DEIM helped project the matrices from a high-dimensional space to a low-dimensional one. The DEIM method has a complexity proportional to the number of variables in the reduced space.

In contrast, POD shows a complexity proportional to the number of variables in the high-dimensional space. Both POD and DEIM allow the collapse of a large number of spatio-temporal correlations. POD and DEIM can work as a sampling method to prepare the RBF network input. In this work, a space-filling strategy has been used to build the realizations. Permeability in one and permeability and injection rate in another model are the control variables of the model.

In 2013 Chen et al. (105) used a non-intrusive ROM to predict the space–time pressure solutions. This method is called the black-box stencil interpolation method (BSIM). Stencil locality is a key assumption made in the work, which enables a significant reduction in the input parameter space and thus allows deduction of the global solution from local mass conservation principles. Based on this assumption, most controls and uncertainties are typically assumed to have local support. Even though POD helps with dimensionality reduction, the computational complexity is still proportional to the dimension of the high-dimensional problem; DEIM is thus used to reduce the dimensions even more. DEIM can be thought of as a sampling method in this approach.

Assuming the saturation solver is available, Chen et al. calculated pressure at each time step using the saturation values. Two approaches were used in this study. In the first, a Laplacian model was used to represent the simplified physics. The pressure difference between the reservoir simulator outputs (true physics) and simplified Laplacian model (simplified physics) at each grid block and time is calculated. An artificial neural network (ANN) model is built with the stencil model properties as input and the aforementioned difference as output. POD and DEIM are used to calculate the reduced pressure values at new time steps. Having the pressure values, simplified physics, and the ANN model, the pressure at each time step and the grid block is calculated. This methodology was performed on a synthetic 2-D, two-phase model with only 900 grid blocks. In the second approach, POD-DEIM is used only as a sampling method, and the pressure values are estimated using an ANN model. The most important deficiencies of this work are that the PDE has been simplified too unrealistically, considering only permeability and pressure in the formulation. Besides, the saturation is assumed to be known at all time steps and is used as an input for calculating the pressure, which is not convincing.

In 2014 Fedutenko et al. (106) used an RBF-based proxy model to predict the cumulative oil production, water injection, and steam-to-oil ratio for the entire field in a steam-assisted gravity drainage (SAGD) process of a synthetic reservoir model (with three pairs of wells). Requiring fewer number of fully fledged simulation runs has been mentioned as the reason behind using RBF. However, the number of simulation runs needed for this relatively small model is between 50 and 200. A close match is not obtained unless 200 training runs are used for proxy model development, which defeats the purpose of building proxies.

In 2000 Güyagüler et al. (107) used neural network-based proxy models as part of a hybrid optimization process to pinpoint the best well positions in the Gulf of Mexico water flooding project. The proposed hybrid optimization technique was based on the genetic algorithm (GA), polytope algorithm, Kriging algorithm, and neural networks. The net present value of the water flooding project was used as the objective function. Their work focused on improving the efficiency of the optimization itself, rather than using speedier evaluation of the objective function. The points evaluated during the progression of the GA were used to estimate the unvisited points in the search space by calibrating and using a proxy. The Kriging-based proxy resulted in better results compared to neural networks. The authors related this to the generalization incapability of the built network.

A workflow for screening/optimization of cyclic pressure pulsing in naturally fractured reservoirs was studied in a work performed by Artun et al. (108). Different proxy models were developed using neural networks for two different gas injection types (CO$_2$ and N$_2$), and an injection scheme. The reservoir characteristics, fluid properties, and design parameters were changed to build the proxy models with an output of production rate, cumulative oil production, number of cycles, and duration of production. The GA was used on the proxy model to find the optimized scenario. Although the method is called universal, it was developed with a single well model having a single layer.

In a similar work in 2012, Parada and Ertekin (109) implemented a multilayer cascade feed-forward back-propagation ANN algorithm to develop proxy models, which help screening for improved oil recovery (IOR) methods. The rock and fluid properties, as well as the design parameters for 200 reservoir simulation runs were used as an input to the proxy model to obtain the expected total recovery and project lifetime. These values help in selecting the best fitting IOR scenario for the provided reservoir. The models should be four, five, seven, or nine spots and symmetric in terms of all their properties. The screening tool is designed based on homogenous reservoir models.

As explained above, many of the proxy models require modification to the reservoir simulation tools, and these models are in the intrusive model category. Nonetheless, commercial simulation tools do not provide open access to the mathematical model, which makes this category of proxy models impractical for use in the industry. In many cases, the available mathematical models used in reservoir simulators may not be sufficient to explain all the physics involved in the phenomena going on under the surface, e.g., production from shale plays, the EOR process, fracturing, and so on. Data-driven models provide the advantage of using the data to convey what is occurring without going into the detailed mathematical formulation of the physics of the phenomenon.

7.8.1 Surrogate Reservoir Models

Surrogate reservoir models (SRMs) are categorized as smart proxy models. SRMs are approximations of the full-field 3-D numerical reservoir models that are capable of accurately mimicking the behavior of the full-field models. Unlike statistically based proxy models, which require hundreds of simulation runs (110–112), SRMs can be created in a few simulation runs (5–20 simulation runs of complex numerical simulation models that represent very large real-world fields). In 2006, the SRM was presented for the first time by Shahab Mohaghegh to solve the problem of time-consuming runs for an uncertainty analysis of a giant oil field with 165 horizontal wells in the Middle East (22). The reservoir simulation model included about one million grid blocks and took 10 hours to run using a cluster of twelve 3.2 GHz parallel processors.

In this study, SRM was used as an objective function for a Monte Carlo simulation to build thousands of simulation runs in a very short time compared to numerical simulators. Mohaghegh describes the SRM thus (24): "SRMs are ensemble of multiple, interconnected neuro-fuzzy systems that are trained to adaptively learn the fluid flow behavior from a multi-well, multilayer reservoir simulation model, such that they can reproduce results similar to those of the reservoir simulation model (with high accuracy) in real-time." Since 2006, applications of SRMs as an accurate and rapid replica of a numerical simulation model have been reviewed in different studies (24).

SRMs are developed using data extracted from the realizations of the simulation model. The data is included in a spatio-temporal database. Building this database is the first step in developing AI-based reservoir models. The main objective of this database is to teach the model the whole process of fluid flow phenomena in the reservoir. Meticulous efforts should therefore be considered in this stage. The quality and quantity of this database determines the degree of success in developing a successful AI-based reservoir model including an SRM. Not dedicating enough attention to this part is the main reason behind unsuccessful attempts at applying AI-based models in the literature (113). Mohaghegh thoroughly discussed this step of SRM development in his paper in 2011 (23).

In order to create the spatio-temporal database, the first step is to identify the number of runs that are required to develop the SRM. The purpose behind having different realizations of a reservoir simulation model is to introduce the uncertainties involved in the model to the SRM. This is a common step in building SRMs and developing response surface methods. However, there is a key difference between these two methods: the functional forms behind the models. Response surfaces are developed using statistical approaches, which use predetermined functional forms. The outputs of reservoir simulation models are then fitted to these predetermined forms. In order to match these functional forms, hundreds of runs are needed.

On the other hand, the pattern recognition characteristics of SRMs help to develop these types of model by having a small number of simulation runs. However, there is no algorithm to find the optimum number of simulation runs to build an SRM. The common practice when choosing the best number to train an SRM is to use rules of thumb based on the intricacy and heterogeneity of the reservoir model, which might change. Nevertheless, it is obvious that if the number of simulation runs is too small, the SRM will not be able to reproduce the simulator results properly. If the number of simulation runs is too big, there is no reason to develop an SRM because the solution replicates the original problem, which is a high number of simulation runs.

After running the realizations, the static and dynamic data are extracted to build the representative spatio-temporal database. The database includes different types of data, such as static and dynamic reservoir characteristics, operational constraints, and so on. Static data refers to properties of the reservoir that are not changing over time, such as permeability, porosity, top depth, and thickness. Dynamic data refers to any data that change over time, such as well constraints or pressure and phase saturation (24).

The training process of an SRM includes three different steps: training (learning), calibration, and validation. Based on this, the spatio-temporal database is divided into three sets: the training or learning set, the calibration set, and the validation or verification set. The training set is part of the data shown to the ensemble of neural networks used during the training process. The neural networks are adapted to this set to match the provided outputs (reservoir simulation results). The calibration set is not used to adjust the outputs. This set is utilized to assure that any increase in accuracy over the training data set will lead to an increase in accuracy over a data set that has not been seen by the neural networks. This set of data is helpful in determining when the training should be stopped.

Finally, the verification set is the part of the database used to verify the predictability of the trained neural networks; consequently, this data set is not used to train the neural networks. It is worth mentioning that the elapsed time to perform the training process (learning, calibration, and verification) is negligible when compared to the reservoir simulation run time. Another important point is that an SRM may be a collection of several neural networks that are trained, matched, and verified in order to generate different results.

A further validation step is utilized in SRM development to assure its robustness. This step is referred to as "blind verification." It is called "blind" because it is a set of realizations that has not been used during the training process. These blind testing sets are complete realizations of the reservoir, while the verification set used in the training process is a randomly selected portion of the spatio-temporal database. In a recently published book, all details associated with generating data-driven reservoir models are presented (114).

7.9 Development of the Smart Proxy

In order to introduce the uncertainties involved in the reservoir model to an SRM (smart proxy), a small number of geological realizations are built and executed using a commercial numerical reservoir simulator. Ten realizations were developed and used to train the smart proxy for pressure and 16 realizations were developed and used to train it for phase saturations. An extra five realizations were generated that were completely independent of the total of 26 realizations used for training and calibration. These five realizations were used as blind simulation runs in order to validate the predictive capabilities of the trained SRM for both pressure and phase saturations.

The properties that were modified in order to generate new realizations for this study included permeability distributions for nine layers of the reservoir (layers 1, 2, 19, 20, 21, 22, 23, 24, and 25) and flowing BHP at 45 injection wells. The reason behind varying the permeability distribution maps for only nine layers goes back to the base model. In the base model, the permeability variation is only noticeable in the named layers while it is consistently low in the other layers. Figure 7.11 shows the permeability distributions for the

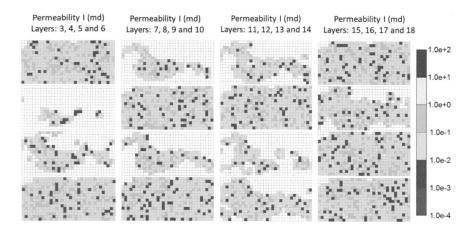

FIGURE 7.11
The permeability distribution for the layers; permeability does not alter through different realizations.

layers, which were not altered during SRM development. In this figure, the low permeability range (less than 1 mD) is notable. To generate the permeability distributions for other layers, the range of permeability in the base model was used. Additionally, the range for varying flowing BHP is 60%–100% of the litho-static pressure.

A "design of experiment" (DOE, also known as "experimental design") method was utilized over the properties' range to construct combinations of the input parameter values such that the maximum information can be obtained from the minimum number of simulation runs. Latin hypercube sampling (LHS) was the experimental design method used in this study. Latin hypercube sampling has enjoyed popularity as a widely used sampling technique for the propagation of uncertainty in analyses of complex systems (103). Using the experimental design method, the range and average of the permeability distribution is constrained to the base model. The distribution of permeability changes over different realizations.

It is assumed that the permeability values at the well locations are available (in reality coming from the core data and its correlation with well logs, as is common practice in the geo-modeling community). Therefore, using a geo-statistical method (inverse distance estimation provided in CMG-Builder), a distribution of permeability can be generated. Figure 7.12 displays the process of generating new realizations while altering the permeability distribution and BHP at injection wells, for the development of the SRM in this project.

The permeability distributions at different layers for training and validation realizations are shown in Figures 7.13 and 7.14. Each row in these figures represents a scenario, and training and validation realizations are indicated. Each column in the figures shows the permeability distribution for a particular layer at different realizations. Figure 7.15 displays the variety of flowing

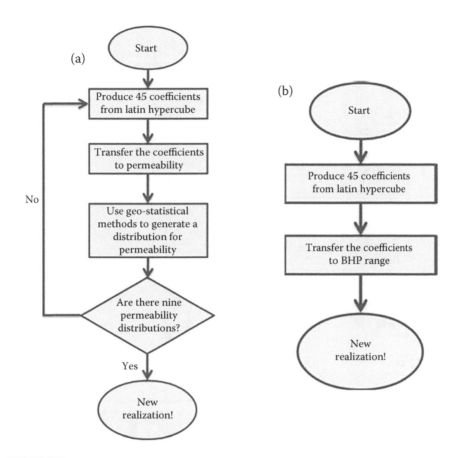

FIGURE 7.12
(a) Flow chart to generate different realizations by altering the permeability distribution.
(b) Flow chart to generate different realizations by altering the flowing bottom-hole pressure
at injection wells.

BHP values at the injection wells used for the realizations incorporated into
the training and validation process. There are 13 scenarios presented in
Figure 7.15. Among these scenarios, numbers 1 to 10 are training realizations,
and scenarios 11 to 13 have been used to validate the final smart proxy.

In the path to develop the SRM, neural networks should be trained,
calibrated, and validated. In order to generate neural networks, IDEA™*
software was used (Figure 7.16). IDEA™ is a software application made for the
development of general data-driven, intelligent models. Figure 7.17 displays
the inputs and outputs of the SRM that were used during the development
(training, calibration, and validation) process. IDEA™ provides a random
data-partitioning algorithm to set the training, calibration, and verification
segments of the data set. As mentioned, the spatio-temporal database was

---

* Intelligent Data Evaluation & Analysis (IDEA™) software is built by Intelligent Solution Inc. (ISI).

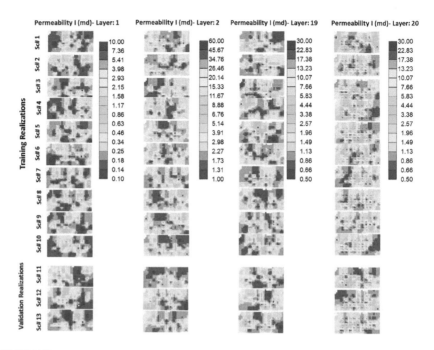

FIGURE 7.13
Permeability distributions at layers 1, 2, 19, and 20 for ten training and three blind realizations.

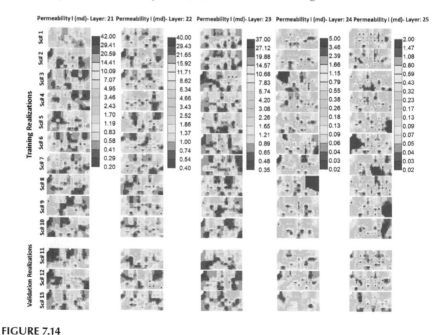

FIGURE 7.14
Permeability distributions at layers 21, 22, 23, 24, and 25 for ten training and three blind realizations.

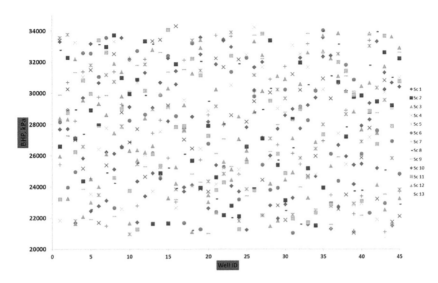

FIGURE 7.15
Flowing bottom-hole pressure at 45 injection wells for ten training and three blind realizations.

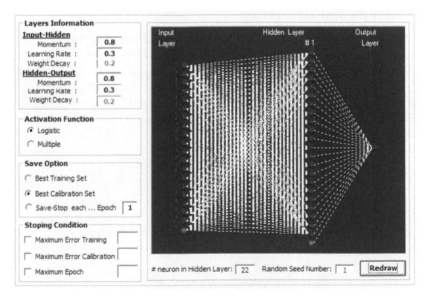

FIGURE 7.16
Structure of artificial neural network built in IDEA™.

built based on information from ten simulation runs. The training, calibration, and verification included 80%, 10% and 10% of the data in the database, respectively. After development of the SRM, its robustness was verified using blind realizations. These runs were not used at any step of training, calibration, or verification. Back-propagation was used as the training algorithm.

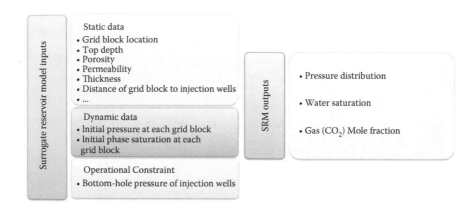

FIGURE 7.17
Inputs and out puts of the SRM. The input parameters include static data, dynamic data, and operational constraints, and the output parameters are pressure, water saturation, and mole fraction.

7.10 Results and Discussion

The SRM was trained, calibrated, and validated using a small number of simulation runs. In these realizations, the distributions of permeability (at nine layers) and flowing BHP for injection wells are the variable properties. In order to validate the robustness of the SRM, it was deployed on blind realizations of the reservoir model. The blind cases of the reservoir simulation models were not used during the training process of the SRM.

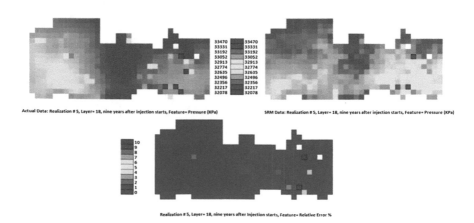

FIGURE 7.18
Comparison between the results of the simulation model (left) and SRM (right) for pressure distribution for a training realization at layer 18, nine years after injection starts. The figure below represents relative error.

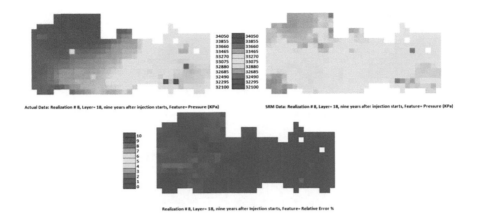

FIGURE 7.19
Comparison between the results of the simulation model (left) and SRM (right) for pressure distribution for a training realization at layer 18, nine years after injection starts. The figure below represents relative error.

In this study, the SRM was trained and validated to reproduce the results of the reservoir simulation model (pressure, water saturation, and CO₂ mole fraction) at the target layer (layer 18) for different time steps during and after injection of the CO₂. Layer 18 is the first layer above the injection layers, and this was chosen to demonstrate the effect of changing the variable parameters on the pressure and phase saturation behaviors in this layer. The motivations behind this study originated from the labor- and time-intensive characteristics of reservoir simulation models. The computational time associated with a single realization of the reservoir simulation model in this study is about 4 to 24 hours

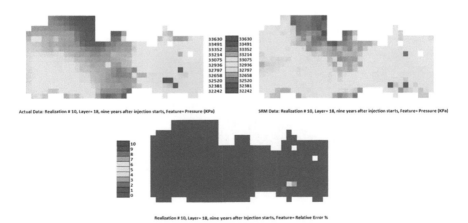

FIGURE 7.20
Comparison between the results of the simulation model (left) and SRM (right) for pressure distribution for a training realization at layer 18, nine years after injection starts. The figure below represents relative error.

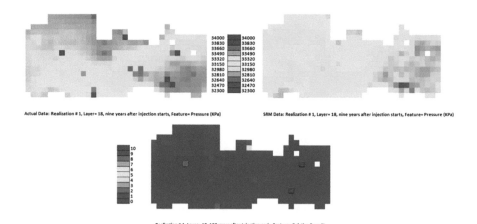

FIGURE 7.21

Comparison between the results of the simulation model (left) and SRM (right) for pressure distribution for a blind (validation) realization at layer 18, nine years after injection starts. The figure below represents relative error.

(depending on convergence time) on a six-processor computer with 24 gigabytes of RAM (random access memory). A typical analysis of a CO$_2$ sequestration problem requires hundreds of realizations. On the other hand, a validated SRM runs in the order of seconds using the same computational power.

In addition to the high speed of the SRM, this AI-based model is able to accurately replicate the results of the reservoir simulation model. The SRM was developed to predict the distribution of pressure, water saturation, and gas (CO$_2$) mole fraction at layer 18 for ten different time steps. In this study,

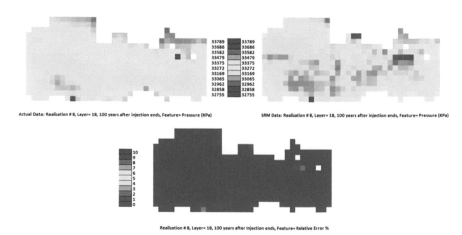

FIGURE 7.22

Comparison between the results of the simulation model (left) and SRM (right) for pressure distribution for a training realization at layer 18, 100 years after injection ends. The figure below represents relative error.

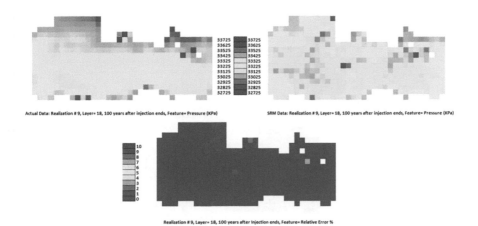

FIGURE 7.23
Comparison between the results of the simulation model (left) and SRM (right) for pressure distribution for a training realization at layer 18, 100 years after injection ends. The figure below represents relative error.

two time steps including injection and post injection periods were selected as representative of the results. These two time steps consist of one injection and one post injection period. The first time step is during the injection period and is nine years after the injection starts. Note that the total amount CO₂ is injected in 50 years. The second selected time step is during the post injection period, and shows the results for 100 years after injection ends in order to detect any change in pressure or saturation that may have taken place due to any possible leakage or other potential phenomena. For each time step, three training realizations and one blind realization were chosen to be displayed in this chapter.

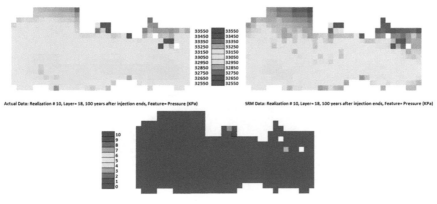

FIGURE 7.24
Comparison between the results of the simulation model (left) and SRM (right) for pressure distribution for a training realization at layer 18, 100 years after injection ends. The figure below represents relative error.

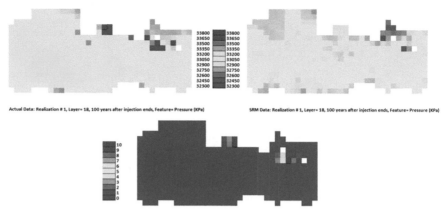

Actual Data: Realization # 1, Layer= 18, 100 years after injection ends, Feature= Pressure (KPa)   SRM Data: Realization # 1, Layer= 18, 100 years after injection ends, Feature= Pressure (KPa)

Realization # 1, Layer= 18, 100 years after Injection ends, Feature= Relative Error %

FIGURE 7.25
Comparison between the results of the simulation model (left) and SRM (right) for pressure distribution for a blind (validation) realization at layer 18, 100 years after injection ends. The figure below represents relative error.

The accuracy of the SRM in reproducing the results of the simulation model is illustrated in Figures 7.18 through 7.41. Figures 7.18 through 7.20 demonstrate the pressure distribution at the target layer (layer 18) during injection (nine years after injection starts) for three different realizations used to train the SRM. These images show the results of the simulator compared to the SRM. The relative error distribution between the simulator and the SRM is shown

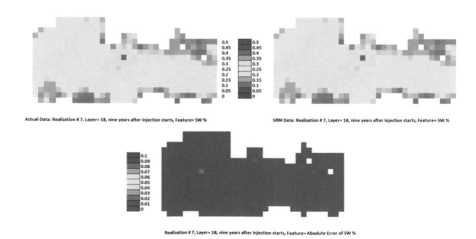

Actual Data: Realization # 7, Layer= 18, nine years after Injection starts, Feature= SW %   SRM Data: Realization # 7, Layer= 18, nine years after Injection starts, Feature= SW %

Realization # 7, Layer= 18, nine years after Injection starts, Feature= Absolute Error of SW %

FIGURE 7.26
Comparison between the results of the simulation model (left) and SRM (right) for water saturation distribution for a training realization at layer 18, nine years after injection starts. The figure below represents absolute error.

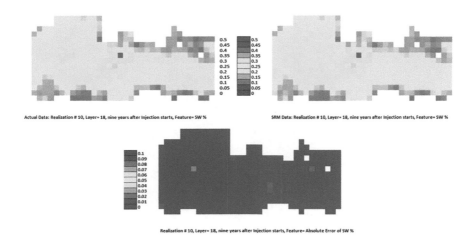

Comparison between the results of the simulation model (left) and SRM (right) for water saturation distribution for a training realization at layer 18, nine years after injection starts. The figure below represents absolute error.

at the bottom of each figure. The SRM predicts the pressure distribution very well, and the relative error distribution confirms this.

There are a few blocks that are out of the normal range due to numerical problems in the simulation model that cause issues with the pressure behavior. Although the SRM understands the general behavior at these blocks, it does not show a similar performance to the other blocks. The reason for this goes back to the pattern recognition characteristics of the SRM: it cannot learn

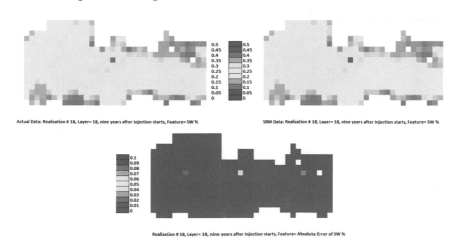

FIGURE 7.28
Comparison between the results of the simulation model (left) and SRM (right) for water saturation distribution for a training realization at layer 18, nine years after injection starts. The figure below represents absolute error.

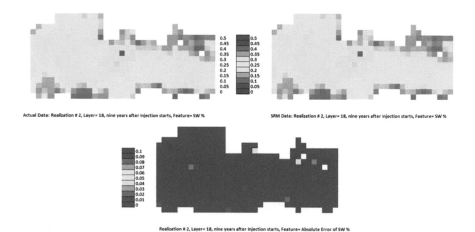

Actual Data: Realization # 2, Layer= 18, nine years after injection starts, Feature= SW %

SRM Data: Realization # 2, Layer= 18, nine years after injection starts, Feature= SW %

Realization # 2, Layer= 18, nine years after injection starts, Feature= Absolute Error of SW %

FIGURE 7.29

Comparison between the results of the simulation model (left) and SRM (right) for water saturation distribution for a blind realization at layer 18, nine years after injection starts. The figure below represents absolute error.

a pattern that is out of the training range. Figure 7.21 presents the results for the same property and time step (pressure distribution for nine years after injection starts) for a blind (validation) scenario. It is obvious that the distribution of pressure is different in different realizations, although they are in a similar range. The main reason for this behavior is alteration of the permeability distribution at the bottom layers (which are injection layers) for different realizations.

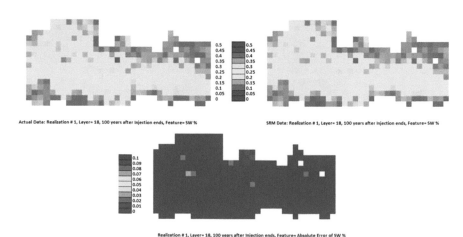

Actual Data: Realization # 1, Layer= 18, 100 years after injection ends, Feature= SW %

SRM Data: Realization # 1, Layer= 18, 100 years after injection ends, Feature= SW %

Realization # 1, Layer= 18, 100 years after injection ends, Feature= Absolute Error of SW %

FIGURE 7.30

Comparison between the results of the simulation model (left) and SRM (right) for water saturation distribution for a training realization at layer 18, 100 years after injection ends. The figure below represents absolute error.

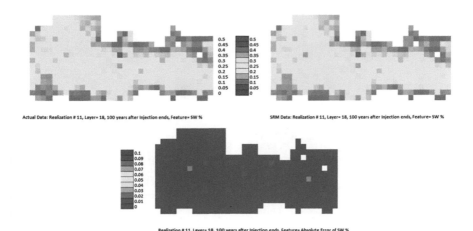

FIGURE 7.31

Comparison between the results of the simulation model (left) and SRM (right) for water saturation distribution for a training realization at layer 18, 100 years after injection ends. The figure below represents absolute error.

Figures 7.22 through 7.24 compare the pressure results for training realizations after 100 years, when the injection plan ends, and Figure 7.25 displays the results for a blind realization. The general relative errors for the pressure distribution are less than 10%.

The results for the water saturation distribution are shown in Figures 7.26 through 7.32. Figures 7.26 through 7.28 display the results of the numerical simulator and the SRM for three different realizations used in training,

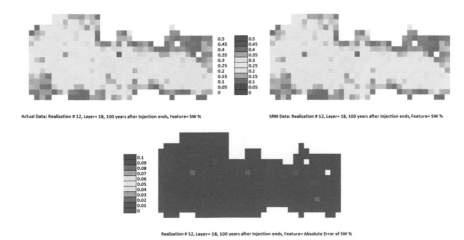

FIGURE 7.32

Comparison between the results of the simulation model (left) and SRM (right) for water saturation distribution for a training realization at layer 18, 100 years after injection ends. The figure below represents absolute error.

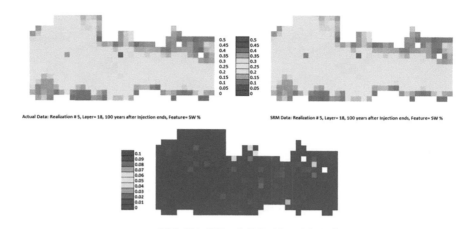

Actual Data: Realization # 5, Layer= 18, 100 years after injection ends, Feature= SW %   SRM Data: Realization # 5, Layer= 18, 100 years after injection ends, Feature= SW %

Realization # 5, Layer= 18, 100 years after injection ends, Feature= Absolute Error of SW %

FIGURE 7.33

Comparison between the results of the simulation model (left) and SRM (right) for water saturation distribution for a blind (validation) realization at layer 18, 100 years after injection ends. The figure below represents absolute error.

calibration, and validation sets. The bottom of these figures shows the absolute error between the simulator and the SRM outputs.

Figure 7.29 demonstrates the same results and absolute error distributions for one blind realization. These figures (Figures 7.26 through 7.29) are the results for nine years after injection starts. Although the changes in the water saturation are not as great as the changes in the pressure (CO$_2$ is the injected fluid and water does not tend to move due to low permeability values at this

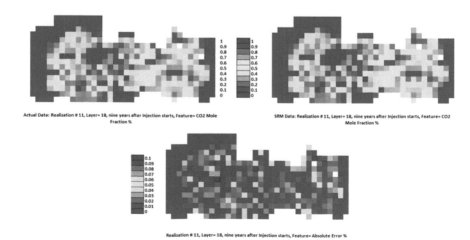

Actual Data: Realization # 11, Layer= 18, nine years after injection starts, Feature= CO2 Mole Fraction %   SRM Data: Realization # 11, Layer= 18, nine years after injection starts, Feature= CO2 Mole Fraction %

Realization # 11, Layer= 18, nine years after injection starts, Feature= Absolute Error %

FIGURE 7.34

Comparison between the results of the simulation model (left) and SRM (right) for gas (CO$_2$) mole fraction distribution for a training realization at layer 18, nine years after injection starts. The figure below represents absolute error.

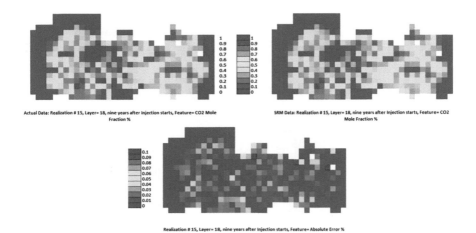

FIGURE 7.35
Comparison between the results of the simulation model (left) and SRM (right) for gas (CO_2) mole fraction distribution for a training realization at layer 18, nine years after injection starts. The figure below represents absolute error.

layer), the SRM performs well in these realizations. Figures 7.30 through 7.33 show the same results for the post injection time step (100 years after injection ends). The general absolute error for this property is as large as 3%.

Figures 7.34 through 7.41 illustrate and compare the results of the simulator and the SRM for the gas (CO_2) mole fraction. Figures 7.34 through 7.37 describe the results and absolute errors of training (three realizations) and blind realizations for a time step during the injection period (nine years after injection

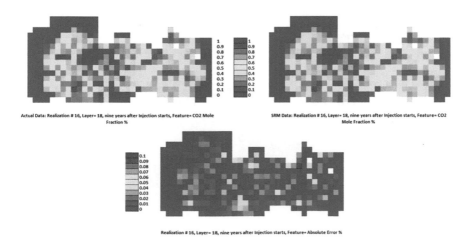

FIGURE 7.36
Comparison between the results of the simulation model (left) and SRM (right) for gas (CO_2) mole fraction distribution for a training realization at layer 18, nine years after injection starts. The figure below represents absolute error.

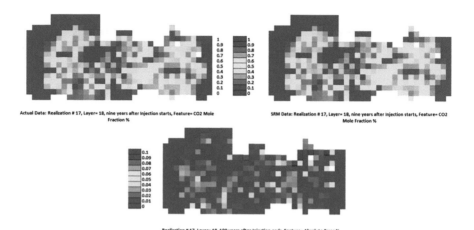

FIGURE 7.37

Comparison between the results of the simulation model (left) and SRM (right) for gas (CO$_2$) mole fraction distribution for a blind (validation) realization at layer 18, nine years after injection starts. The figure below represents absolute error.

starts). Figures 7.38 through 7.41 show the same property for a post injection time step (100 years after injection ends). Although the general absolute error for the gas mole fraction increases to 10%, the results of the SRM are satisfactory.

Figures 7.18 through 7.41 prove the accuracy of the developed SRM in this study. The number of simulation runs required to train the SRM was surprisingly low. When compared to the computational power and the time needed for running the simulation model, the SRM shows its efficiency.

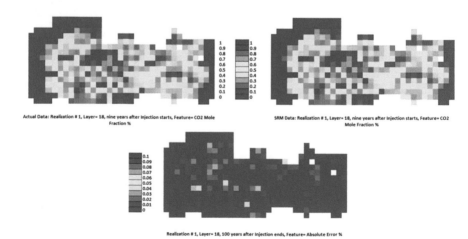

FIGURE 7.38

Comparison between the results of the simulation model (left) and SRM (right) for gas (CO$_2$) mole fraction distribution for a training realization at layer 18, 100 years after injection ends. The figure below represents absolute error.

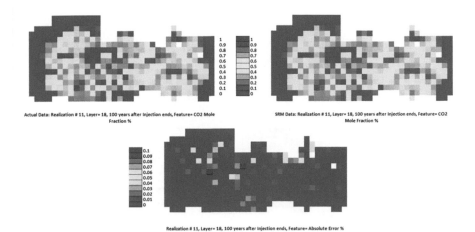

FIGURE 7.39
Comparison between the results of the simulation model (left) and SRM (right) for gas (CO$_2$) mole fraction distribution for a training realization at layer 18, 100 years after injection ends. The figure below represents absolute error.

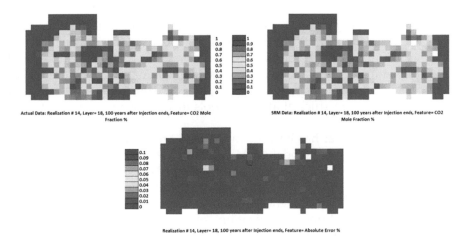

FIGURE 7.40
Comparison between the results of the simulation model (left) and SRM (right) for gas (CO$_2$) mole fraction distribution for a training realization at layer 18, 100 years after injection ends. The figure below represents absolute error.

7.11 Concluding Remarks

The consequences of the daily increasing CO$_2$ concentration in the atmosphere have been shown as a real threat for life on this planet. Carbon capture and storage (CCS) has shown potential as a practical method to reduce the amount of CO$_2$ originating from human activities in the atmosphere. In order to secure

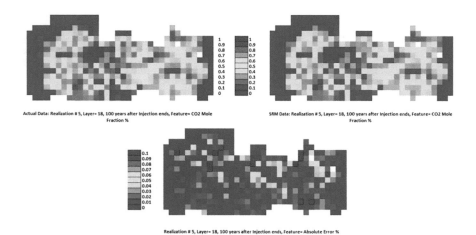

Actual Data: Realization # 5, Layer= 18, 100 years after Injection ends, Feature= CO2 Mole Fraction %

SRM Data: Realization # 5, Layer= 18, 100 years after Injection ends, Feature= CO2 Mole Fraction %

Realization # 5, Layer= 18, 100 years after Injection ends, Feature= Absolute Error %

FIGURE 7.41
Comparison between the results of the simulation model (left) and SRM (right) for gas (CO$_2$) mole fraction distribution for a blind (validation) realization at layer 18, 100 years after injection ends. The figure below represents absolute error.

the stability of a CCS project, a comprehensive study of fluid flow through porous media is required. The conventional tools used to perform such an analysis are numerical reservoir simulation models. Although numerical reservoir simulators are able to perform a detailed analysis, they are highly time-consuming and computationally expensive. Pattern-recognition-based reservoir models are efficient alternate tools to address the aforementioned issues.

The technology that has been demonstrated and utilized here is known as a surrogate reservoir model (SRM). SRM is the application smart proxy modeling in reservoir simulation and modeling. The capabilities of an SRM to be a fast and accurate replica of a reservoir simulation model make it an efficient tool to perform conventional analyses in the petroleum industry.

In this study, ten different realizations of the base model were designed to develop the SRM to predict pressure behavior in a reservoir. Sixteen realizations were considered in order to simulate the phase saturation behavior. The comprehensive spatio-temporal database was developed based on the data extracted from these realizations. The SRM was trained, calibrated, and validated using a data-driven and intelligent model developer software. The robustness of the SRM was further validated using blind realizations.

8

Leak Detection in CO$_2$ Storage Sites

Shahab D. Mohaghegh and Alireza Haghighat

CONTENTS

The ability of underground CO$_2$ storage to confine and sustain injected CO$_2$ for a very long time is the main concern regarding geological CO$_2$ sequestration. If leakage from a geological sink occurs, it is crucial to find the approximate amount and location of the leak in order to implement proper remediation activity.

An overwhelming majority of research and development for storage site monitoring has concentrated on atmospheric, surface, or near surface monitoring of the sequestered CO$_2$. This chapter proposes developing an in situ CO$_2$ monitoring and verification technology based on the implementation of permanent down-hole gauges (PDG) or "smart wells" along with artificial intelligence and data mining (AI&DM). The technology attempts to identify the characteristic of the CO$_2$ leakage by de-convolving the pressure signals collected at the smart well sites. Since creating an actual leak in an existing CO$_2$ storage site is not a viable option for testing this technology, the demonstration of this technology's predictive capabilities is performed through numerical simulation and modeling of an actual CO$_2$ storage site in the Citronelle field.

In Chapter 5 of this book, the development and history matching of a numerical reservoir simulation model for CO$_2$ sequestration in the Citronelle field was covered in detail. The presence of PDGs is considered in the reservoir model at the injection well and an observation well. High-frequency pressure data from sensors is collected based on different synthetic CO$_2$ leakage scenarios in the model. Due to the complexity of the pressure signal behaviors, a machine learning-based technique is introduced to build a real-time intelligent leakage detection system (RT-ILDS).

The RT-ILDS is able to detect leakage characteristics in a short time (less than a day), and demonstrates high precision in quantifying the leakage characteristics subject to complex rate behaviors. The performance of the RT-ILDS is examined under different conditions such as multiple well leakages, cap rock leakage, availability of an additional monitoring well, presence of pressure drift and noise in the sensors, and uncertainty in the reservoir characteristics.

8.1 CO_2 Leakage from Underground Storage

As mentioned previously, carbon capture and storage (CCS) is considered to be the ideal short-term strategy for sustaining or reducing the global atmospheric CO_2 concentration. The technology for capturing and transporting CO_2 from producing sources like power plants, petrochemicals, cement, metal and other factories to the sinks (underground geological storages) has been widely used in the chemical and petroleum industry. In the storage aspect of CCS, complications are arising as the operations are relatively new and need to be investigated thoroughly in order to determine if the geological sinks are suitable for the long-term storage of CO_2. Notably, the potential for CO_2 leakage from the underground storage to the atmosphere is an important issue that needs to be addressed.

The sinks for geological CO_2 sequestration are depleted petroleum and gas reservoirs, deep saline aquifers, and coal beds. Leakage in underground CO_2 storage will undermine the benefits of the geological storage of CO_2. Also, leakage could have harmful ecological impacts and present health risks above and beyond the issues related to global warming and climate change. As far as leakage from CO_2 storage sites is concerned, leakage rate performance requirements have been established to be at or less than about 0.1% annually (116). To meet these standards, CO_2 storage sites must have active monitoring systems to detect CO_2 leakage, and must be prepared to take remedial actions in the event of leakages. In order to select an appropriate monitoring system, adequate knowledge of the characteristics of the leakage (its approximate amount and location) is required.

8.1.1 CO_2 Leakage Conduits

8.1.1.1 Well Leakage

The source, driving force, and pathway are the three most probable causes for the development of a leakage in an underground storage. The leakage source is the injected CO_2. The driving force for CO_2 movement can be considered as the buoyancy or particularly the pressure difference between the source and surface (or leakage location in the reservoir) due to the injection. In the presence of the source and driving force, the well bore can be a pathway (Figure 8.1) if it includes a poorly cemented casing, casing failure, or abandonment failure (117). These pathways may have been present at the well bore prior to the CO$_2$ injection. Also, after injection, it is possible that CO$_2$ causes cement degradation and casing corrosion.

Wells are especially important because they provide a direct and almost vertical pathway through the formations that otherwise act as a seal for the injected CO$_2$. Well logs (especially sonic) examination is a good tool for describing the potential of leakage from a well bore. Abandoned wells have a higher probability of providing a pathway for CO$_2$ leakage. Since 2003, regulations require that all the surrounding permeable zones be isolated or covered to prevent any communication between the storage and geological formations (117). When the down-hole cement plugs have been installed, it is necessary to keep the well open for inspection for 5 days. When the well is checked for a fluid level test or other signs of leakage (such as bubbles in the fluid), the top of the casing is cut and capped almost 1 m below ground level (117).

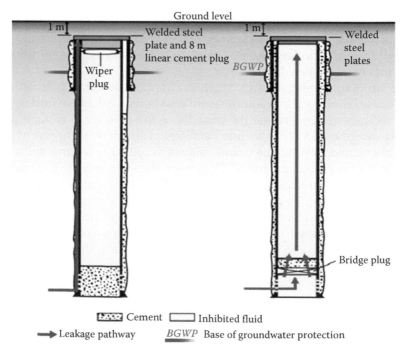

FIGURE 8.1
Different leakage pathways along the well bore. (From Watson, T. L. and Bachu, S. 2007. Evaluation of the Potential for Gas and CO$_2$ Leakage along Wellbores. *E&P Environmental and Safety Conference*. Galveston, TX: s.n.)

In many cases in wells that were abandoned before 2003, the wells were constructed with a low annular cement top allowing cross flow behind the casing. With current regulations, a cement squeeze is needed to achieve good isolation, in conjunction with putting some inhibitor liquids inside the casing. In addition, the pressure should be increased to 7,000 kPa and casing vent flow tests should be performed to maximize efficiency (117).

If any flow of gas is observed, a repair process has to be carried out before abandonment. The test for surface casing vent flow (SCVF) is referred to as a situation where the pressure in the casing or annula is sustained, indicating that gas has entered the production casing from a formation. Wells with positive SCVF that exhibit gas flow rates greater than 300 m³/day or have a stabilized build-up pressure of more than 9.8 kPa/m must be repaired immediately (117). The pressure build up in the SCVF can be used to determine the properties and characteristics of the leaked well, especially the effective permeability of the leak. This is done by assuming a continuous Darcy flow for the CO_2 movement along the well's leak path. In Figure 8.2, based on casing pressure build-up observations, the leak effective permeability was calculated to be 1.18E-20 m².

Another test that is required in some regulations is gas migration (GM). The GM test is based on creating 50 cm deep holes in the soil around the wellbore directed outward. These holes should be sealed and let the possible leaked gas to accumulate. To detect combustible leaked gas, frequent readings of

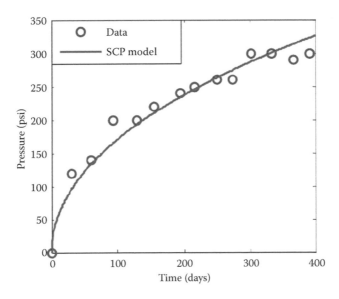

FIGURE 8.2
Best fit to observed (in a field) casing pressure data, resulting in leak permeability estimation. (From Tao, Q. et al. 2010. *Model to Predict CO₂ Leakage Rates Along a Wellbore.* Florence, Italy: Society of Petroleum Engineers. ISBN: 978-1-55563-300-4 (319).)

TABLE 8.1

Factors That Impact Potential Well Leakage

Factors Showing No Apparent Impact	Factors Showing Minor Impacts	Factors Showing Major Impact
Well age	License	Geographic area
Well operational mode	Surface casing depth	Well bore deviation
Completion interval	Total depth	Well type
H$_2$S or CO$_2$ presence	Well density	Abandonment method
	Topography	Oil price, regulatory changes
		Uncemented casing/hole annulus

Source: Watson, T. L. and Bachu, S. 2007. Evaluation of the Potential for Gas and CO$_2$ Leakage along Wellbores. *E&P Environmental and Safety Conference*. Galveston, TX: s.n.

lower explosion limits (LEL) should be made. Further investigation should be performed in an event of gas detection (117).

In a data-mining study for about 316,500 wells in the province of Alberta in Canada, various factors were investigated to determine if the potential for leakage could be assessed based on well information. The factors are categorized in brief in Table 8.1.

8.1.1.2 Cap Rock Leakage

Prior to CO$_2$ injection, existing cap rock or confining unit, seals the reservoir and prevent any communication with any other permeable media above or below the reservoir. After injecting CO$_2$ and pressuring the reservoir, the pressure difference among target zone and the other side of caprock increases. Once the pressure difference across the cap rock exceeds minimum horizontal stress, confining unit (seal) may be fractured generating flow paths to the lower pressure layers (51).

When the seal is breached, initially the fracture conductivity across the cap rock is very low. But pressure difference across the confining unit may result in fluid flow along the fracture. This fluid flow can increase the fracture conductivity and impact the seal quality. As the pressure difference and outflow decreases, fractured closure may occur. At this point, damaged seal can heal temporarily, however seal can be breached much easier along the fracture plane (51).

In order to determine the cap rock leakage characteristics, the stress distribution in the reservoir and cap rock should be determined by geo-mechanical models or measurements. These models can also orientate the minimum principal stresses which are most prone to be fractured and provide a leakage path for CO$_2$ (119). The best result for leakage characterization will be achieved if coupled geo-mechanics and flow models are used. The reservoir simulator computes pressure and temperature, which are used as inputs for the geo-mechanical models in order to determine the stress distribution and consequently the rock failure and leakage permeability.

Fracture permeability is described as the permeability that occurs due to rock breaching during pressurizing of the reservoir by CO$_2$ injection. If the cap rock crack finds a way to an overlying permeable layer, CO$_2$ can escape from the reservoir. The Barton-Bandis (119) model is generally used to demonstrate the fracture permeability behavior in a reservoir. Based on this model, no fracture exists in the matrix before the pressure increase starts. Another assumption in this model is high brittleness of the rock resulting in a maximum value for the permeability at the beginning of the fracturing. From the beginning, the fracture aperture remains open until the pressure in the rock drops (119). This pressure reduction leads to a decrease in the fracture aperture and consequently the leakage permeability, as demonstrated in Figure 8.3.

Another factor that prevents a cap rock acting as a seal is the displacement of connate water in the pores or fractures due to buoyancy forces. In other words, the capillary entry pressure of the largest interconnected pore throat must be bigger than the pressure exerted to the cap rock by CO$_2$ (buoyancy). The difference between water and CO$_2$ densities as well as the height of the CO$_2$ column determine the magnitude of the buoyancy force. The factors determining the magnitude of the resistant force are rock wettability, the largest connected pore throat radius, and the gas–water interfacial tension (52). By applying the force balance, the cap rock seal strength corresponds to the height of the CO$_2$ column that can be retained in the reservoir:

$$\text{Critical height} = \frac{2\gamma \times [1/r_t - 1/r_p]}{g \times [\rho_w - \rho_{CO_2}]}$$

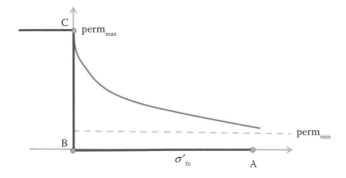

FIGURE 8.3
The Barton-Bandis model. (From Tran, D. et al. 2009. Geomechanical Risk Mitigation for CO$_2$ Sequestration in Saline Aquifers. *SPE Annual Technical Conference and Exhibition*. New Orleans, LA: SPE.)

where

    $\gamma =$ interfacial tension
    $r_t =$ radius of pore throats
    $r_p =$ radius of pores
    $g =$ acceleration due to gravity
    $\rho_w =$ water density
    $\rho_{CO2} = CO_2$ density

8.1.1.3 Fault Leakage

Faults are considered as potential pathways that result in CO_2 migration from the target formation into the atmosphere or other subsurface formations. Due to the existence of faults in most sedimentary basins, a fault–fluid interaction evaluation should be given more attention, especially for a CO_2 storage risk assessment. Two important parameters are involved in fault evaluation: "fault sealing capacity" and the "fault region petro-physical description" (120). The fault seal capacity indicates if a fault is acting as a barrier to flow. If the fault is non-sealed, its conductivity can be considered by the fracture or matrix permeability, which limits the flow but does not stop it. The seal capacity of a fault can be quantified by the shale gouge ratio (SGR). SGR can be defined as an estimate of the amount of shale in the fault based on the averaging or mixing rule shown in Figure 8.4:

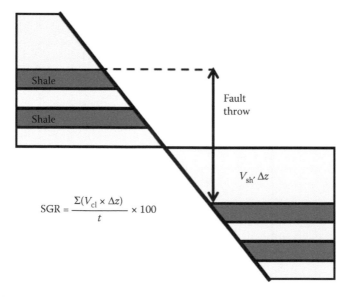

FIGURE 8.4

Shale gouge ratio calculation. (From Chang, W. C. 2007. *A Simulation Study of Injected CO$_2$ Migration in the Faulted Reservoir.* Austin, TX: The University of Texas at Austin.)

$$SGR = \frac{\sum \Delta z_i Vs_i}{Throw} \times 100\%$$

where
   Vs_i = shale concentration in layer i
   Z_i = thickness of layer i

Another parameter that affects fault conductivity is membrane sealing, which is actually the minimum capillary entry pressure that must be overcome before a non-wetting phase (CO_2 in this case) can enter into the fault's pore throat (the fault is mainly filled with ground particles). Understanding these two parameters helps to quantify how long it takes for buoyant CO_2 to migrate from the fault region to the atmosphere and how much CO_2 leaks through the fault (120). The main fault characteristics are presumed to be functions of lithology (clastic or carbonate), fault structure, and sealing mechanisms. Due to the difficulty and practicality issues (it is preferred to keep the integrity of the fault) associated with sampling (core) fault rocks in deep formations, fault property specification is challenging.

8.1.2 CO₂ Leakage Impacts

Usually, two types of risk are associated with the underground storage of CO_2: the risk of CO_2 leakage through the paths to the overlying formations during and after the injection process (discussed in the previous section), and the risk of aquifer over-pressurizing during injection. When CO_2 is injected into saline formations, it creates a pressure build up that may lead to damage to the seal, such as fracturing or fault activation, or brine leakage out of the reservoir. The CO_2 leakage risk (CLR) depends on the probability of CO_2 leakage and its consequences and impacts:

$$CLR = \text{Total probability} \times \text{Impact}$$

The impact of CO_2 leakage can be assessed by its flow rate and concentration, as higher flow rates have a more severe impact. The total probability of the leakage is divided into three separate probabilities (36). The first probability (1) considers the chance that the CO_2 plume intersects with existing leakage paths (faults, wells) in the reservoirs. The second probability (2) deals with the likelihood of a path connection to other compartments or the atmosphere. The final probability (3) takes into account the chance of conductivity or sealing of the leakage path:

$$\text{Total Probability} = \text{Probability (1)} \times \text{Probability (2)} \times \text{Probability (3)}$$

CO_2 leakage may result in serious effects on humans, animals, and the ecosystem. CO_2 is heavier than the O_2 in the air, so its leakage and release results in high accumulations in cellars and valleys, with serious consequences. CO_2 leakage into shallow ground levels may contaminate and affect the quality of water, soil, and mineral resources. An increase in CO_2 concentration in water decreases its pH, which leads to an increase in the hardness of water (120).

8.2 Storage Site Monitoring

There are several techniques for monitoring the storage site during the injection process as well as after completion of injection.

8.2.1 Well Monitoring

CO_2 injection wells can be considered the most probable leakage path in underground storage sites. Wells are supposed to keep their integrity over the injection period and then during the post injection time for an estimated 10,000 years. Integrity refers to the safe operation of the well throughout its service life by reducing the unintended risk of CO_2 release. Well monitoring provides a preventive verification process to see if the integrity of the well is being maintained. A leak detection log (LDL), tubular inspection, production (injection) profile, neutron, spectral and cement bound logs can be used as cased-hole logs for the validation of cement integrity, pressure isolation, corrosion, and injection profiles (35).

8.2.1.1 Permanent (Pressure) Down-Hole Gauges (PDGs)

Reservoir pressure is probably the most reliable parameter for understanding subsurface flow behavior. PDGs have been used in the oil and gas industry for decades to measure well pressure over different time periods, and to analyze reservoir behavior and characteristics. The advancement of this technology enables operators to install permanent DHGs, which transduce real-time and instantaneous data. Real-time pressure data can be used to detect fluid movement in the reservoir, which is an indicator for CO_2 movement and migration (35).

8.2.2 Seismic Imaging

The petroleum industry has been using seismic waves for decades to obtain subsurface images for underground geological modeling and interpretation. In this technique, sound waves are emitted into the subsurface. The reflected waves are altered and contain key information about the subsurface rock geometry, distribution, properties, and boundaries.

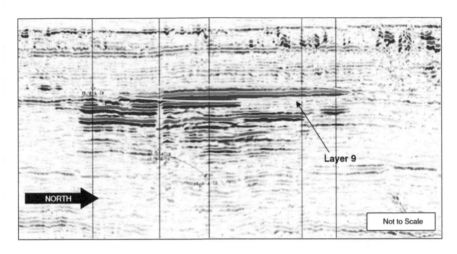

FIGURE 8.5
Seismic monitoring of CO_2 injection in Sleeper Field. (From Singh, V. P. et al., 2010. *Reservoir Modeling of CO₂ Plume Behavior Calibrated Against Monitoring Data From Sleipner, Norway.* Florence, Italy: Society of Petroleum Engineers. ISBN: 978-1-55563-300-4 (78).)

Recently, advances in interpretation tools have resulted in the use of time-lapse seismic images (4-D seismic) to monitor CO_2 movement in underground storage sites. In this method, initial seismic images for an area of interest are compared with seismic images taken in the same area at different times.

Movement of CO_2 in the subsurface makes changes in the pore/fluid properties that can be observed by time-lapse seismic analysis (35). The use of 4-D seismic imaging for CO_2 monitoring in depleted gas or low-porosity reservoirs is not practical because the sound waves' responses to CO_2 movement are not detectable within such formations. The 4-D seismic images taken from Sleipner Field are good practical examples of CO_2 movement in an underground storage site, as shown in Figure 8.5.

8.2.3 Gravity Surveys

Changes in the vertical columns' rock densities may be calculated by gravity measurements. Due to the displacement of saline brine within the subsurface by CO_2, a reduction in average column density may occur, and by implementing gravity measurements as a monitoring tool, it is possible to show where the CO_2 leakage is occurring. Gravity surveys are ideally applicable for shallow reservoirs with high porosity and thickness. Unfortunately, the detection of CO_2 movement in reservoirs with porosity less than 10%, thickness less than 10 m, and depths more than 2,500 m is typically not possible by gravity measurements. For practical CO_2 monitoring, fixed gravimeters with high accuracy of about 5 microgals are needed (35).

8.2.4 Satellite Imaging

Currently, the use of satellite platforms to measure vertical ground elevations in different time periods is a viable option. Maps of surface deformation over time are generated based on returned microwave energy analysis. In the petroleum industry, some studies from satellite image observations have shown ground-level subsidence or uplift due to oil production or gas injection, respectively. Prior to use of this monitoring technique, a ground uplift response of 1 mm based on the corresponding volume of injected CO$_2$ should be determined and calibrated. Deviation of ground level uplifts from calibrated values may be an indicator of CO$_2$ leakage (35).

8.3 Strategy for Leakage Prevention and Remediation

To select a safe and secure site for geological storage, leakage pathways, cap rock integrity, and assured natural confinement must be considered along with assured well bore integrity and sufficient reservoir capacity. Of these components, emphasis should be placed on long-term well integrity at the CO$_2$ storage site. In descending order of importance, a series of reservoir simulation-based modeling should be conducted to track and project the movement of the CO$_2$ plume. The purpose of the overall CO$_2$ monitoring system at the storage site installation is to serve as an early warning system and provide online information about the movement and immobilization of the CO$_2$ plume.

By developing a pre-established mitigation strategy, a rapid response would be available on detecting a leakage. Reducing the pressure in the storage formation, increasing the overlying formation pressure, or re-injecting the CO$_2$ in more secure formations are other possible options to stop a leakage (121).

8.4 Smart Well Technology

The concept of the "smart field," which is also recognized in the industry by names such as i-field, e-field, field of the future, digital oilfield, and so on, is a new technical area that is rapidly gaining support and recognition in the oil and gas industry. A smart field includes multiple smart wells. Smart well technology is mainly based on the installation and incorporation of down-hole measurement tools that will make it possible to control many operations related to the well bore, from a distance.

Advances in the technology involving drilling and completion enable the installation of PDGs, which are capable of operating in harsh environments for extended periods of time. PDGs collect and transmit high-frequency data streams to remote offices to be analyzed and used for reservoir management.

8.4.1 Definition of a Smart Well

Smart wells are generally utilized with equipment capable of performing down-hole measurements and/or controlling the production process at the reservoir level. A list of equipment or technologies that can be added to conventional wells in order to convert them to smart ones is shown in Figure 8.6.

8.4.2 Application of Smart Wells

Several reservoir management applications exist with the ability to control and measure variables at the reservoir level. In co-mingled or stacked pay zones, well production cannot reach its optimal value due to differences in reservoir pressure at different compartments. Utilization of down-hole chokes would allow the reservoir to produce from multiple layers with minimum cross flow or fluid loss (123). Horizontal wells also benefit considerably from smart well technology, especially in thin oil rims. Thin reservoirs may experience early water/gas breakthrough, which may be managed by the installment of inflow control valves at different well locations to shut off unwanted flows. Additionally, secondary recovery mechanisms, such as gas/water injection, can be controlled optimally by smart wells to avoid excessive injected fluid production through highly permeable zones (123).

Down-hole measurements also provide the capability of flow profiling by using distributed temperature sensing fiber optics. Through fiber optics, down-hole measurements detect cross flows and flows behind pipes. This is especially practical in wells where production profiling is expensive. Another

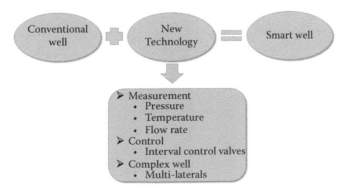

FIGURE 8.6
Smart well instruments and technologies. (From Brouwer, D. R. 2004. *Dynamic Water Flood Optimization with Smart Wells Using Optimal Control Theory*. Delft: Deflt University of Technology (122).)

application of smart wells is the "auto gas lift," where oil-producing wells cross different compartments with an active gas cap (123). Inflow control valves make it possible to use and to control gas from other layers and flow the oil based on artificial lift procedures. One future application for smart wells will be down-hole production testing. PDGs transduce the well flowing pressure and flow rate data, collectively forming the fundamental information used for well test analysis. In addition, down-hole geophones may be installed in the well system, enabling operators to perform repeatable seismic tests to obtain the reservoir imaging data used in monitoring sweep efficiency.

8.5 Closed-Loop Reservoir Management

The high-frequency data streams can be used for real-time monitoring, simulation/model updating, and finally optimal control of oil and gas reservoirs. The combination of all the mentioned processes results in "closed-loop reservoir management," as shown in Figure 8.7. Data from the sensors can be assimilated into the simulator to update and history match the reservoir models. Real-time data indicates if the actual performance of the field is deviating from planned targets. If deviation occurs, appropriate

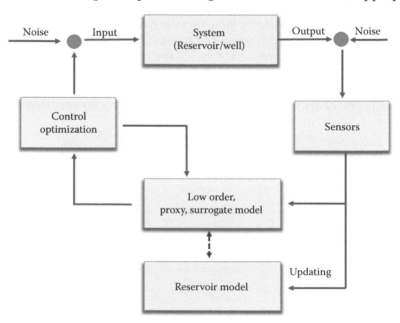

FIGURE 8.7
Closed-loop reservoir management. (From Jansen, J. D. et al. 2009. Closed Loop Reservoir Management. *SPE Reservoir Simulation Symposium*. Woodlands: SPE.)

control actions would be necessary, based on the optimization algorithms, to bring back performance toward targets.

In an underground CO_2 storage it is possible to place an array of PDGs along the injection wells in the formation where the CO_2 is being injected. Notably, gauges may be placed in the injection and/or production wells or in the slim holes drilled specifically for this purpose. Pressure changes in the reservoir can be used during the injection to update the reservoir model and after injection for reservoir characterization and reservoir monitoring purposes (124).

8.6 CO₂ Leakage Detection Using Smart Well Technology: Case Study

CO_2 leakage detection in storage sites using pressure data from PDGs is a fairly new topic in the CO_2 sequestration research area. Several authors (125–130) have tried to investigate this topic with different methodologies. Generally, most of the presented methods attempted to use analytical solutions to find pressure behavior subject to the CO_2 leakage characteristic and solve the inverse problem to find leakage components (128,129). Another methodology, introduced by Jalali et al. (130), used neural networks to find seepage in a CO_2 sequestration model in a coal-bed with multiple sensors (PDGs). All mentioned studies used synthetic models which were completely homogenous with at most two reservoir layers. The significance of the study that is being presented in this chapter over previous works is the use of a history-matched reservoir simulation model developed for a real CO_2 sequestration project (Citronelle Field). Additionally, CO_2 leakage was detected based on a novel data-processing method, implemented for the analysis of real-time pressure data. Finally, the robustness of the method and workflow presented in this chapter is evaluated by considering various reservoir and CO_2 leakage characteristics. Two different CO_2 leakage detection techniques based on smart well technology are discussed briefly in the next subsections.

8.6.1 Leakage Detection: Leakage Test with Analytical Model

A hydraulic or water injection test can be applied to underground saline aquifers in order to examine if the reservoir has the proper confinement capacity for CO_2 storage. For this case study (128), water is injected into the aquifer and the pressure is monitored at the observation well in an upper aquifer for a specific period of time (Figure 8.8).

The objective here is to determine leakage placement and transmissibility based on pressure data. This test is somewhat similar to the pressure interference well test, which is widely used in the petroleum industry to

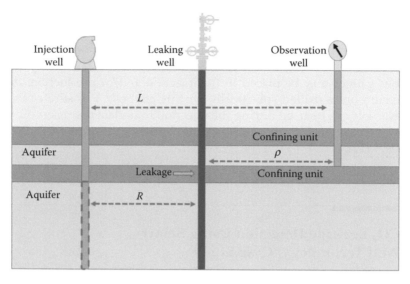

FIGURE 8.8
Leakage test configuration. (From Zeidouni, M. and Pooladi-Darvish, M. 2010. Characterization of Leakage through Cap-Rock with Application to CO2 Storage in Aquifers – Single Injector and Single Monitoring Well. *Canadian Unconventional Resources and International Petroleum Conference.* Calgary: s.n)

determine directional permeability and other major reservoir properties like skin factor and average reservoir pressure. The leakage problem in this case is an inverse problem since the leakage parameters are manipulated so that the modeled pressure matches the measured data. In this test, the analytical solution, which is the dimensionless pressure response at the monitoring well, can be expressed as follows (the bar sign means the equation is given in the Laplace domain (128):

$$\bar{P}_{mD} = \frac{\bar{q}_{lD}(s)K_0(\sqrt{(s/\eta_D)}\rho_D)}{T_D r_{lD}\sqrt{(s/\eta_D)}K_1(\sqrt{(s/\eta_D)}r_{lD})}$$

where
q = flow rate (m^3/s)
l = leak
K_0 = zero-order Bessel function
s = Laplace transform
ρ = leak monitor distance (m)
T = transmissibility (m^3)
D = dimensionless
η = diffusivity (m^2/s)
K_1 = first-order Bessel function
r = radius (m).

When the transient pressure measurement at the monitoring well is given, the inverse problem would be finding the leakage characteristics (cartesian coordinates, leak radius, and permeability) in a way that the difference between the measured pressure (Y) and predicted pressure ($\overline{P_{mD}}$) is minimized:

$$f(x) = Y - \overline{P_{mD}}$$

This method can provide little information about the leakage location (the solution of the objective function with respect to leak location is not unique), but the leak transmissibility can be evaluated within a narrower confidence interval. Different test strategies can enhance the leak characterization capabilities. These strategies include increasing the pressure sampling frequency, the use of pulsing in the water injection, increasing the number of monitoring and/or injection wells, and using a pressure derivative (with respect to time) analysis.

8.6.2 Leakage Detection with Neural Networks and Reservoir Simulation Model

In this case study (130) a horizontal, single-layer, homogenous coal-bed CO_2 storage model with constant reservoir properties (such as permeability, porosity, and thickness) is considered. Sixteen pressure sensors are located at equal spacing throughout the reservoir, as shown in Figure 8.9.

Daily pressure data was recorded from all 16 sensors along with the location of each sensor. Leakage was introduced to the reservoir by creating

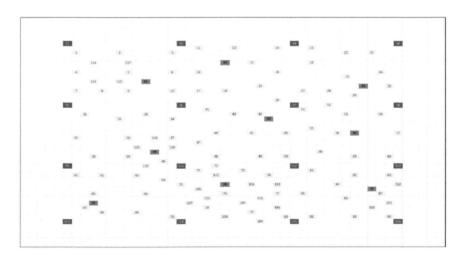

FIGURE 8.9
Leakage location for training and test cases. The 16 pressure sensors are shown in red. (Adapted from Jalali, J. 2010. *Artificial Neural Networks for Reservoir Level Detection of CO₂ Seepage Location Using Permanent Down-Hole Pressure Data*. Morgantown, WV: West Virginia University.)

a small pressure difference, ΔP, between the reservoir pressure and bottom-hole pressure of an imaginary well as the source for CO$_2$ seepage. In this homogenous model, 92 simulation cases were generated.

In each of the cases, a specific location for CO$_2$ seepage was selected. In order to avoid a high reservoir simulation run time, the surrogate reservoir model (SRM) (smart proxy) was used to instantly reproduce the results of the numerical reservoir simulation mode. Definitions of the smart proxy and SRM have been provided in previous chapters of this book, and will not be repeated here. An important feature of SRMs is their fast analysis and generation of outputs in a very short period of time.

For each run, pressure data was collected on a daily basis. The results of these models generate a large data set. The pressure differences between the actual field pressure distributions, recorded by the pressure sensors, and the SRM predictions at the pressure sensors' locations (with no leakage) are then sent to the artificial neural network (ANN) trained for CO$_2$ seepage location detection. The network looks for changes in pressure measurements at the sensors. Once the pressure change exceeds a threshold, it starts searching for the possible location of the seepage. In the case of the CO$_2$ seepage, the ANN provides an approximate location and the amount of CO$_2$ being leaked. Figure 8.10 shows an interface that was developed for this study.

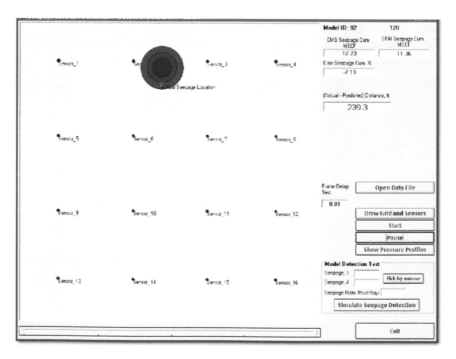

FIGURE 8.10
Computer interface for the neural network prediction of the leakage location.

This study shows that an ANN trained for a heterogeneous reservoir (heterogeneity in porosity and permeability) could detect the location of the seepage with reasonable accuracy, for an area as small as 0.6 acres, in a reservoir with a total area of around 579 acres. The seepage rate in such a reservoir was around 0.3% of the total stored gas per year, which was slightly above the 0.1% per year limit found in the literature.

8.7 Intelligent Leakage Detection System

To assure cap rock integrity, CO_2 storage sites must have active monitoring systems to detect CO_2 leakage and be prepared to take remedial action in the event that leakage occurs. Industry has much experience with a combination of monitoring techniques for underground geological sites, selected primarily based on accessibility and geological characteristics. As already mentioned, the monitoring methods can be classified into two different categories: surface monitoring and underground monitoring and measurements.

In surface monitoring activities, the presence of CO_2 emitted from the ground can be directly assessed (131). Also, CO_2-related parameters like changes in ground-level (elevation) or high-frequency electromagnetic (EM) waves (132) are subject to frequent measurements. Satellite-based optical methods, gas sampling, EM and gravity survey are considered to be types of surface or near surface monitoring. For subsurface monitoring and measurements, the main focus is on tracking the CO_2 plume at the reservoir level. Well logs (pulsed neutron, RST), 4-D seismic, borehole gravity, cross-well seismic, brine–gas composition sampling, and introduced tracers have been applied to monitor the underground movement of CO_2 (42).

Although these methods have been deployed in the field, there are still some drawbacks associated with the practical application of CO_2 monitoring systems. In the surface monitoring method, the main concern is that it is necessary for the CO_2 to be detectable at the surface. Before that time, even though the leakage could have occurred, it would not be possible to detect it (133). Regarding underground monitoring systems, it is worthy to mention that since most of these methods are implemented periodically, it is not possible to detect any leakage during the time interval that no test or monitoring is offered. Therefore the remediation activity and response to the leakage is considered to be reactive with some time lag. This suggests the need for a real-time or online monitoring system in order to detect the CO_2 leakage as fast as possible and thus to provide much more efficient CO_2 leakage risk management.

PDGs and valves have been used to continuously monitor pressure, temperature, flow rates, and automatic flow controls (124). This technology can be used in underground CO_2 reservoirs to monitor pressure in real time. To help accommodate CO_2 leak detection, two PDGs were installed in the

Citronelle Dome's observation wells. A reservoir simulation model for CO$_2$ sequestration at Citronelle Dome was developed (see Chapter 5). In the study presented in this chapter, multiple scenarios of CO$_2$ leakage are modeled, and high-frequency pressure data from the PDGs in the observation well is collected. The complexity of the pressure signal behavior and the reservoir model makes the use of an inverse solution of analytical models impractical. Therefore an alternate solution was developed for an intelligent leakage detection system (ILDS), based on machine learning.

In order to investigate the proof of concept for ILDS, a simple reservoir simulation model of CO$_2$ injection in the Citronelle Field was initially used for CO$_2$ leakage modeling. This model presented homogenous porosity and permeability in every sand layer. After successful deployment of ILDS with a simple and homogenous reservoir model, a history-matched reservoir model (explained in Chapter 5 of this book) was used to build and test an upgraded ILDS. All the steps required for the development of an ILDS with homogenous and heterogeneous reservoir simulation models are covered in detail in the following sections.

8.7.1 ILDS Development Based on the Homogenous Model

8.7.1.1 Reservoir Simulation Model

A reservoir model was built using a commercial numerical reservoir simulator based on interpreted geophysical well logs. The geological model of the Paluxy Formation based on the homogenous model consists of 51 simulation layers. This model includes $50 \times 50 \times 51 = 127{,}500$ Cartesian grids (Δx and Δy equal to 122 m; local grid refinement was applied around the injection well). Based on an initial core study taken from the characterization and monitoring well, constant values for porosity and permeability were assigned to each layer (Table 8.2). The temperature of the reservoir is 383 K. The brine salinity and density values are 100,000 ppm and 993 kg/m^3, respectively. The pressure reference in this model is 30.3 MPa at 2,870 m (true vertical depth [TVD]).

The initial reservoir simulation runs showed that the maximum extent of the CO$_2$ plume is located in the first (top) layer. This is mainly due to the fact that the top layer comprises sand with high permeability, which causes CO$_2$ to migrate further from the injection well. As is shown in Figure 8.11, the

TABLE 8.2

Porosity and Permeability Values for Different Layers in Citronelle Reservoir Simulation Model

Layer	1	2	3	4	5	6	7	8	9	10	11	12	13	14	15	16	17
Porosity (%)	19.8	18	18	19.3	21.8	19.3	18.2	17	18	16	15.5	19.3	19.3	19.3	19.3	19.3	17.5
Permeability (m²*e-14)	43	17	17	9	122	9	19	10	21	6	5	9	9	9	9	9	13

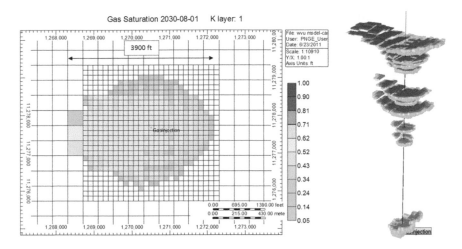

Gas Saturation 2030-08-01   K layer: 1

FIGURE 8.11
Plume extent in the first layer (left) and all layers 25 years after injection (SI conversion: 1 ft = 0.3 m).

approximate diameter of the plume area in the first layer reaches 1,189 m, 25 years after injection has stopped.

8.7.1.2 CO_2 Leakage Modeling

Development of the ILDS is a data-driven technology by nature. Therefore, it requires a training process through which the ILDS learns the fluid flow behavior due to CO_2 leakage in the reservoir, and its impact on pressure changes that occur in the porous media. To accommodate this necessity, the developed numerical reservoir simulation model is used to generate the required data to train the ILDS. The generated data includes high-frequency streams of pressure data that are received from PDGs. To generate this data, it is necessary to design a set of simulation runs that provide the pressure behavior in the observation wells (D-9-8) with respect to different leakage rates and locations (synthetic or artificial leakage).

In the numerical simulation model, nine wells are located in the area of investigation in order to accommodate the different leakage scenarios that need to be modeled. They include the injection well (D-9-7), and eight other wells to be used, when needed, to model the different amounts of leakage at different locations in the field. These wells are named D-9-1, D-9-2, D-9-3, D-9-6, D-9-8, D-9-9, D-9-10, and D-9-11. There are two pressure gauges in well D-9-8. The maximum CO_2 plume extent (1,158 to 1,220 m) is in the first layer. Therefore leakage from these wells can be initiated by perforating them in the first layer.

The focus so far has been on the different leakage rates shown in Table 8.3. These are the rates observed in real cases all around the world (133). Based on analysis of data from the leaking wells, the majority of leaks from the

TABLE 8.3

Leakage Rates Observed in Real Cases

t/year	ft³/day	m³/day
35	1,837.50	52
100	5,250	149
210	11,025	312
800	42,000	1,189
1,400	73,500	2,080
1,900	99,750	2,823
2,300	120,750	3,417
2,500	131,250	3,714
10,000	525,000	14,858
100,000	5,250,000	148,575

Source: Lozzio, M. et al. 2010. *Quantifying the Risk of CO_2 Leakage through Wellbores.* New Orleans, LA: SPE.

well bore are negligible, with limited consequences. These leakage rates were assigned to the existing well bores in order to learn the type of pressure behavior they initiate in a reservoir like the one being modeled in this study.

Based on the above-mentioned study, leakage rates of the order of 35–100 t/year are considered small leakages. Rates from 100 to 800 t/year occur typically in underground CO_2 storage projects. About 93% of wells have a leak rate smaller than 1,400 t/year. A major event, which may result in fatalities and extreme damage, will have a leak rate of the order of 10,000–100,000 t/year.

In order to generate high-frequency pressure data in the observation and injection wells, 20 different CO_2 leakage rates (in the range of real leakage rates observed in actual cases, as shown in Table 8.3) were assigned to wells D-9-2, D-9-6, and D-9-10 (Figure 8.12). These CO_2 leakage rates are shown in Table 8.4.

For each CO_2 leakage scenario, it is assumed that leakage starts 2 years after the end of injection (January 1, 2017). For each CO_2 leakage rate, a reservoir simulation run is performed on an hourly basis (each time step is considered to be one hour). The duration of leakage is 6 months. We also performed a simulation with no leakage to be used as a baseline for comparison. In this case, the bottom-hole pressure in the observation well begins to increase during injection until it reaches its maximum at the end of injection. After the end of injection, the pressure drops until the reservoir reaches equilibrium. The typical time for this period is about 4 to 5 months. After this, the bottom-hole pressure becomes almost constant.

When a CO_2 leakage occurs in one of the wells, it creates a pressure change in the reservoir. This pressure change can be observed in the observation well. The difference between the pressure in the observation well, when no leakage exists and when a leakage happens is considered the leakage indicator (Figure 8.13).

This pressure difference (Δp) behavior can characterize the specifications of the leakage, specifically the two most important characteristics related to

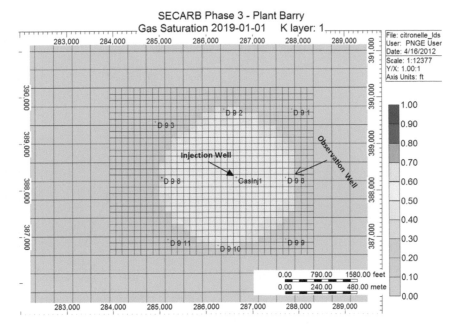

FIGURE 8.12
Location of the wells in the plume extent (SI conversion: 1 ft = 0.3 m).

the leakage: its location and the amount of CO_2 being leaked. For example, the magnitude of Δp is directly proportional to the CO_2 leakage rate. The trend in Δp as a function of time is related to the location of the leakage. As an example, the Δp trend (high-frequency hourly basis) in the observation well for the case where well D-9-6 leaks at a rate of 850 m³/day is shown in Figure 8.14. For all the leakage rate scenarios in Table 8.4 (three wells leak individually), high-frequency Δp values were generated for the observation well.

TABLE 8.4

Leakage Rates Assigned for Wells D-9-2, D-9-6, and D-9-10

Leakage Rate			
t/year	m³/day	t/year	m³/day
286	425	1238	1840
381	566	1333	1981
476	708	1429	2123
571	849	1524	2264
667	991	1619	2406
762	1132	1714	2547
857	1274	1810	2689
952	1415	1905	2830
1048	1557	2000	2972
1143	1698	2095	3113

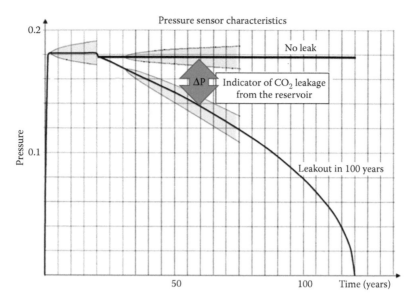

FIGURE 8.13
Reservoir pressure behavior during the leakage.

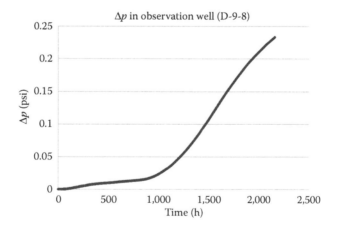

FIGURE 8.14
Δp in the injection well in the case where the D-9-6 well leakage rate is 850 m³/day (SI conversion: 1 psi = 6.9 kPa).

8.7.1.3 Data Summarization

Normally, the data transmitted from the PDG sensors can be categorized as noisy high-frequency data streams. The first step in processing such data streams is to remove the noise associated with the data. The process of de-noising will be explained comprehensively in the following sections. The high-frequency PDG data should be summarized and transformed into a format that can be used by the pattern recognition technology.

Based on the characteristics of Δp (high-frequency data streams), "descriptive statistics" was used for data summarization. Descriptive statistics quantitatively describes the main features of a collection of data and provides simple summaries about the sample and about the observations that have been made (134). These summaries may form the basis of the initial description of the data that will be used by the pattern recognition technology.

The parameters that can represent and summarize a large amount of data are as follows: mean, standard error, median, mode, standard deviation, sample variance, kurtosis, skewness, range, maximum, minimum, and sum. For example, the descriptive statistics and summarization of 504 hourly Δp data points in the observation well (D-9-8) for the case where well D-9-6 leaks at a rate of 3,115 m^3/day are presented in Table 8.5.

Another way to represent large numbers of data points that can also be implemented in the neural network training and pattern recognition is through curve fitting. In this process, a curve is constructed, providing a mathematical function that has the best fit to a series of data points. The trend of the Δp history curve is different with respect to the location of the leakage. Therefore it is not possible to determine a typical curve (linear, exponential, and so on) to fit all the data points. The only curve that can provide a good fit for Δp points is the "polynomial" curve. In this study a 4th degree polynomial curve was used to fit Δp points for different leakage rates and leakage locations.

TABLE 8.5

Descriptive Statistics and Summarization of 504 Hourly Δp Data Points in Observation Well D-9-8

Descriptive Statistics	
Mean	0.053
Standard error	0.0034
Median	0.012
Mode	0
Standard deviation	0.077
Sample variance	0.0059
Kurtosis	1.088
Skewness	1.52
Range	0.28
Minimum	0
Maximum	0.28
Sum	26.7
Count	504

For example, for the case where well D-9-6 leaks at a rate of 850 m³/day (Figure 8.14), the following polynomial mathematical function represents the best fit with $R^2 = 0.9992$:

$$\Delta p = -7e\text{-}14t^4 + 3e\text{-}10t^3 - 3e\text{-}07t^2 + 0.0001t - 0.0085$$

The coefficients and intercept of the mentioned mathematical relation can be used for pattern recognition and neural network training. Based on neural network training results, using descriptive statistics parameters leads to much more accurate predictions compared with a network trained with coefficients of the fitted curve. Therefore descriptive statistics would be used from now on for data summarization.

8.7.1.4 Data Partitioning for Neural Network Modeling

In order to construct a neural network model, it is first necessary to prepare a data set including input and output features. In this study, the aim is to determine the location and amount of leakage, based on data provided by PDGs. Therefore, the latitude and longitude (X, Y) of the leaking well (D-9-2, D-9-6, and D-9-10) and the CO_2 leakage rates are the output features of the neural network. The CO_2 leakage rates are shown in Table 8.4.

The actual input data received directly from the PDGs is in the form of pressure readings from the observation well or, in other words, the difference between pressure readings during leakage and "no leaking" conditions (Δp). As explained in the previous section, Δp readings at different times (hourly basis) are summarized into a set of descriptive statistics parameters. For the initial study, pressure information (PDG readings) in the observation well after 1 week of leakage (hourly basis) was selected.

Prior to input data selection, key performance indicator (KPI) analysis should be completed in order to determine which parameters are more influential to be considered as inputs. The first KPI test was performed on the location of the leakage or the coordinates of possible leaking wells. Results of the KPI analysis for the location of leakage are shown in Figure 8.15.

Rank	Feature	% Degree of Influence	Rank	Feature	% Degree of Influence
1	Skewness	100	1	Skewness	100
2	Kurtosis	89	2	Kurtosis	44
3	Range	12	3	Standard Error	15
4	Standard Deviation	11	4	Range	15
5	Standard Error	11	5	Maximum	15
6	Sum	11	6	Standard Deviation	14
7	Sample Variance	9	7	Sum	14
8	Maximum	8	8	Sample Variance	9
9	Mean	6	9	Mean	2
10	Median	1	10	Median	1

FIGURE 8.15
Key performance indicator analysis results for the latitude (left) and longitude (right) of the leakage.

Rank	Feature	% Degree of Influence	
1	Skewness	▓▓▓▓▓▓▓▓▓▓▓	100
2	Kurtosis	▓▓▓▓▓▓	57
3	Maximum	▓▓	19
4	Standard Deviation	▓▓	19
5	Mean	▓▓	18
6	Sum	▓▓	18
7	Range	▓▓	18
8	Sample Variance	▓▓	15
9	Standard Error	▓▓	15
10	Median		1

FIGURE 8.16
Key performance indicator analysis results for the CO_2 leakage rate.

Based on the results of the KPI analysis, skewness, kurtosis, and range (or maximum) have the most influence on the location of the leakage, and mean and median have the least degree of importance. Therefore the median was not selected as input data for neural network training.

A similar analysis was performed to determine the effect of the parameters on the CO_2 leakage rate. According to the results of the KPI analysis shown in Figure 8.16, skewness, kurtosis, and maximum have the greatest effect on the CO_2 leakage rate, and the median has the least effect. According to the results of the KPI analysis, it was decided to select 10 inputs (mean, standard error, mode, standard deviation, sample variance, kurtosis, skewness, range, maximum, and sum) for neural network training. In this case we assigned 20 different CO_2 leakage rates to three leakage locations (wells D-9-2, D-9-6, and D-9-10). As a result, a total of 60 different records (each record representing a scenario) including 10 input parameters for each scenario were considered for neural network training. For this data set, intelligent data partitioning was used for the segmentation of the records, with 80% of data allocated to neural network training, 10% for network calibration, and 10% for validation. Therefore 48 records were used for training, 6 for calibration, and 6 for validation.

8.7.1.5 Neural Network Architecture Design and Results

Error backpropagation is one of the most popular learning algorithms used in the training of supervised neural networks. To train neural networks for the ILDS, a backpropagation training algorithm was used. For this, just one hidden layer was provided. Based on 10 inputs and 3 outputs, 12 neurons in the hidden layer and one random seed number were allocated to the neural network (Figure 8.17). The random numbers initialize the weights on the neural network prior to training.

As shown in Figure 8.17, there are two sets of synaptic connections in the ANN. The first comprises the synaptic connections between the input layer

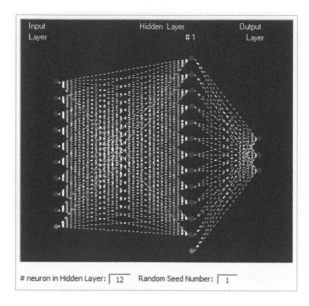

FIGURE 8.17
Architecture of the neural network used for ILDS.

and the hidden layer, and the second comprises those between the hidden layer and the output layer. Since we used "Vanilla" and "Enhanced" versions of the backpropagation algorithms for the training of the neural networks, for each connection set, two parameters – momentum and learning rate – were assigned. "Learning rate" determines how fast the network learns the information presented. This is usually a moderate to low number (between 0 and 1). A small learning rate may prolong the learning process and slow it down. Momentum is an extra push to the learning process that serves two purposes. First, it may accelerate the learning process, and second it has the potential to kick the solution out of the potential local minimum, which usually exists in the search space and causes the solutions to converge prematurely (135). In this project, learning rate and momentum were considered to be 0.3 and 0.8, respectively. Also, a logistic activation function was used to connect the input layers to the hidden layers.

The next step is to identify how and when to save the trained network. It is recommended to save the network that has achieved the best training set, or the calibration set. Also, the network can be saved after one epoch or more of training. Each epoch of training is completed when all the records in the training set have been visited by the network once.

The initial results of the neural network training are illustrated in the plots in Figure 8.18. These plots compare actual data (leakage rate and location) with neural network predictions. The neural network quantifies the location of the leaking well with precise accuracy ($R^2 = 1$). For leakage rates, the neural network results cannot predict some of the actual data correctly ($R^2 = 0.92$),

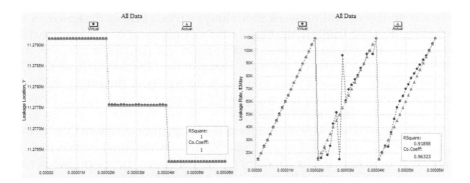

FIGURE 8.18
Actual leakage locations and corresponding neural network predictions (SI conversion:
1 ft = 0.3 m) (left) and actual leakage rates and corresponding neural network predictions
(SI conversion: 1 ft^3 = 0.028 m^3) (right).

specifically the rates belonging to well D-9-6. In order to improve the
results for CO_2 leakage rate predictions, a neural network was individually
developed for each leaking well. This approach was successful in enhancing
the predictive performance of the neural network model for the leakage rate
(Figure 8.19). The R^2 for the data presented in this figure is 0.96.

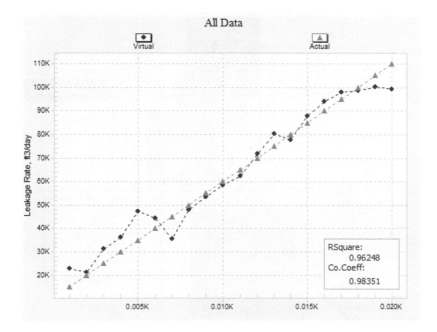

FIGURE 8.19
Neural network prediction for leakage rate, network trained for each well individually
(SI conversion: 1 ft^3 = 0.028 m^3).

Based on the neural network modeling results, ILDS is designed in the following manner. Initially, the high-frequency pressure data is summarized by descriptive statistics, and then summarized features of the pressure data are fed to the main neural network that predicts the location of the CO$_2$ leakage. Afterwards, when the location is determined, the pressure data is fed into the corresponding neural network designed for that specific location. This workflow is shown in Figure 8.20.

In order to validate the performance of the ILDS, three different CO$_2$ leakage rates – 736, 1,472 and 2,492 m^3/day (26, 52, and 88 Mcf/day), which had not previously been seen by the neural network (data from these simulation runs was not used during the training, calibration, and validation of the neural networks) – were assigned to a possible leakage location (wells D-9-2, D-9-6, and D-9-10) as blind validation runs.

Pressure data from these runs was summarized by descriptive statistics and fed into the ILDS. The ILDS predictions for CO$_2$ leakage location and rate are shown in Table 8.6 and Figure 8.21, respectively. The prediction by the ILDS for leakage location is highly accurate, and the results are almost the same as the actual values. For leakage rate predictions, the results are also similar to the actual values, although for the low leakage rate (26 Mcf/day or 736 m^3/day), they differ minimally with the actual values, but the range of predicted rates is reasonably correct.

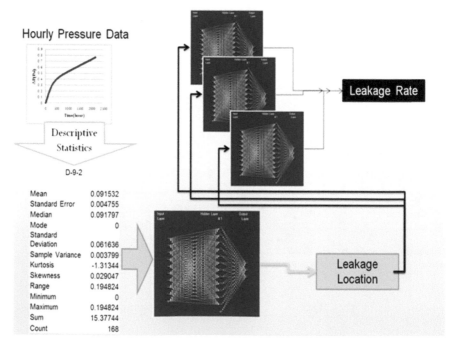

FIGURE 8.20
Workflow for the ILDS.

TABLE 8.6

Actual Leakage Locations and ILDS Predictions

Run	Leakage Location, X Actual (m)	Leakage Location, X ILDS (m)	Leakage Location, Y Actual (m)	Leakage Location, Y ILDS (m)
1	386,761	386,762	3,437,402	3,437,403
2	386,761	386,762	3,437,402	3,437,402
3	386,761	386,761	3,437,402	3,437,403
4	387,206	387,205	3,437,887	3,437,887
5	387,206	387,205	3,437,887	3,437,887
6	387,206	387,205	3,437,887	3,437,887
7	387,152	387,152	3,436,992	3,436,993
8	387,152	387,152	3,436,992	3,436,993
9	387,152	387,152	3,436,992	3,436,993

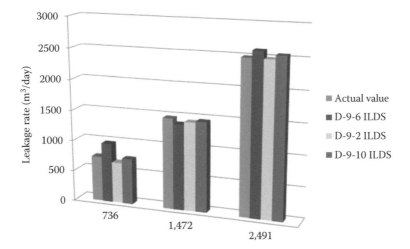

FIGURE 8.21
ILDS leakage rate predications.

8.7.2 ILDS Development Based on the Heterogeneous Model

In this section, development of an ILDS system using a heterogeneous, history-matched model is presented. The availability of additional actual field data resulted in updating the reservoir simulation model explained in Chapter 5 of this book. The history-matched model represents a more realistic situation (CO$_2$ storage project) than the homogenous model, by considering the porosity data obtained by comprehensive well log interpretation and permeability–porosity correlation. Additionally, the assimilation of real field data led to the modification of some reservoir parameters to simulate a pressure value closer to actual measurements. The updated reservoir simulation was used to model high-frequency pressure signal behavior subject to various CO_2 leakage scenarios.

8.7.2.1 CO_2 Leakage Modeling

Five different wells are located in the area of interest in this reservoir. The area of interest is defined by the extent of the CO_2 plume. This area is indicated in yellow in Figure 8.22. Each well can be a possible leakage path in the absence of proper well integrity. Since wells D-9-7#2 (the injection well) and D-9-8#2 (the observation well) were drilled recently, specifically for CO_2 storage purposes, the probability of CO_2 leakage through these wells will be negligible due to the fact that the operator will be well aware of the regulatory requirement and will be mindful of the leakage possibility. Therefore, in these analyses, the focus regarding CO_2 leakage and the detection of its rate and location will be on the remaining three wells.

These are wells D-9-6, D-9-7, and D-9-8. It is assumed that these wells may be vulnerable to some sort of leakage. As discussed in the previous sections of this chapter, when a leakage occurs, signals representing pressure change (Δp) are observed in the observation well. The pressure change signals in the observation well for a leakage rate of 1,840 m^3/day (65,000 ft^3/day) at well D-9-7 are illustrated in Figure 8.23.

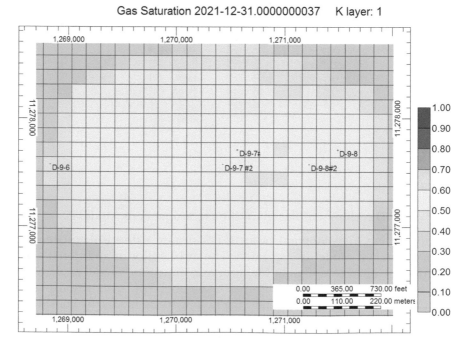

FIGURE 8.22

CO_2 plume extent in the history-matched model (SI conversion: 1 ft = 0.3 m).

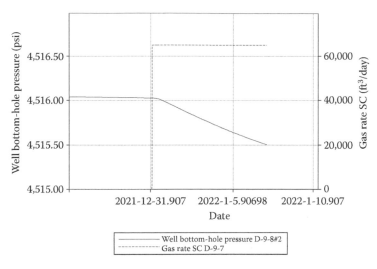

FIGURE 8.23
Reservoir pressure in observation well D-9-8 (SI conversion: 1 psi = 6.9 kPa, 1 ft^3 = 0.028 m^3).

8.7.2.2 Neural Network Data Preparation

An ILDS consists of a neural network that has learned patterns that correlate the leakage characteristics (location and rate) to the corresponding pressure signals. In order to train the neural network, it is necessary to have pressure signals corresponding to each leakage scenario. Reservoir simulation was used to generate the mentioned pressure signals. Each well that was prone to leakage (wells D-9-6, D-9-7, and D-9-8) experienced different leakage rates in the range of 425–2,973 m^3/day with 283 m^3/day increments (15,000–105,000 ft^3/day with 10,000 ft^3/day increments). The simulated leakage initiates at January 1, 2022 and 168 pressure signals per hour are recorded from the observation well (D-9-8#2). Descriptive statistics was used to summarize the pressure signals and to develop the data set to be used for training, calibration, and validation of the neural network. Intelligent data partitioning was used to divide the data set into training (80%), calibration (10%), and validation (10%) portions. The data set was analyzed by KPI in order to identify the relative impact of each input parameter with regard to the summarized pressure signals that serve as the output of the neural network. The results of the KPI analysis are shown in Figure 8.24.

8.7.2.3 Neural Network Architecture and Results

The back-propagation learning algorithm was used to train the networks using all the parameters that were analyzed during the KPI analyses as input parameters. Leakage locations (x coordinate) and leakage rates were set as the output parameters. Based on input–output selection, 10 hidden neurons within one hidden layer formed the structure of the neural network

Rank	Feature	% Degree of Influence	Rank	Feature	% Degree of Influence
1	Skewness	100	1	Kurtosis	100
2	Kurtosis	76	2	Skewness	45
3	Standard Error	21	3	Sample Variance	7
4	Range	20	4	Standard Error	6
5	Maximum	18	5	Sum	6
6	Sum	16	6	Mean	6
7	Standard Deviation	16	7	Maximum	5
8	Mean	15	8	Median	2
9	Sample Variance	12	9	Range	2
10	Median	1	10	Standard Deviation	1

FIGURE 8.24
Key performance indicator for leakage location (left) and leakage rate (right).

(Figure 8.25). Input layers were fully connected to the hidden layers by a logistic activation function. Also, one random seed number was used to start initialization of the neural network weights.

In order to validate the ILDS that was developed by heterogeneous and history-matched model, nine different simulation runs were performed to generate pressure signals corresponding to the leakage rates that were not seen by the neural network during the training process. This is called a completely blind validation test. Three different leakage rates were assigned to each well as shown in Table 8.7, and the pressure signals were collected and summarized by descriptive statistics. The results of the blind validation

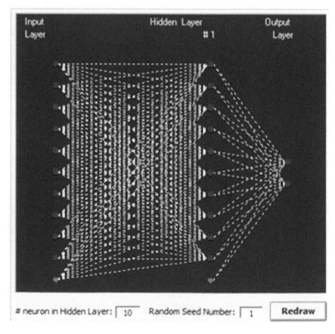

FIGURE 8.25
Neural network architecture for the heterogeneous reservoir model.

TABLE 8.7

Leakage Rates and Locations for ILDS Validation

Run Number	Leakage Rate (m³/day)	Leakage Location (m)	Well
1	651	386,739	D-9-6
2	2,038	386,739	
3	2,632	386,739	
4	906	387,267	D-9-7
5	1,726	387,267	
6	2,462	387,267	
7	764	387,552	D-9-8
8	1,358	387,552	
9	2,858	387,552	

tests are shown in Figures 8.26 and 8.27. In these figures, results generated by the neural networks were compared with actual data (generated by the simulation runs) to investigate the predictive capabilities of ILDS for the scenarios that it has never seen before.

The precision of the neural networks predictions can be analyzed by looking at the error plots and R^2. As shown in Figure 8.28 (left), R^2 for the leakage location predictions is 0.99, which shows high precision in the neural network predictions. Additionally, the error for the leakage location prediction ranges between −1.5 and 2.1 m (actual leakage locations are 386,739, 387,267, and 387,552 m). For the results of the leakage rates (Figure 8.28, right), R^2 is

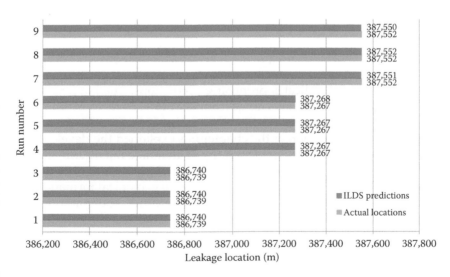

FIGURE 8.26
Results for neural network validation–leakage locations.

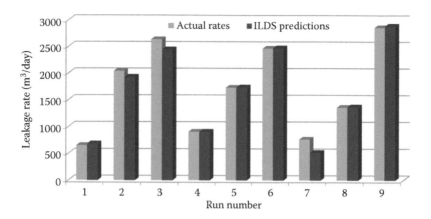

FIGURE 8.27
Results for neural network validation–leakage rates.

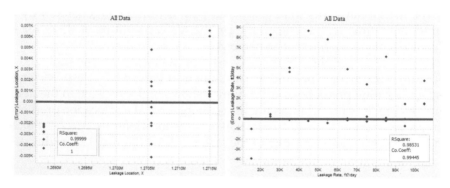

FIGURE 8.28
Neural network prediction errors for leakage location (left) and leakage rates (right).

equal to 0.98, which indicates promising predictions. The maximum errors for the leakage rate predictions range from −141 to +254 m^3/day, which is less than 10% of the actual rates.

In summary, this section of the chapter has explained the development of the next generation of intelligent techniques that take maximum advantage of data collected using "smart field" technology to continuously and autonomously monitor and verify CO$_2$ storage in geological formations. This technology will provide the means for in situ detection and quantification of CO$_2$ leakage in the reservoir. Injection of CO$_2$ in a saline reservoir (Citronelle Dome) was modeled and studied in order to predict reservoir performance, specifically under a variety of modeled CO$_2$ leakage scenarios. CO$_2$ leakage was modeled considering the existence of PDGs in the observation well. High-frequency pressure data was processed and summarized by descriptive statistics. Finally, an ILDS was designed, developed, and blind-tested with simple homogenous as well as complex history-matched, heterogeneous reservoir models.

The main findings of this chapter may be summarized as follows:

- The pattern recognition capabilities of artificial intelligence and data mining (AI&DM) may be used as a powerful de-convolution tool.
- Locating the CO_2 leakage points and quantifying CO_2 leakage rates in storage sites, using "smart field" technology, is a technologically feasible concept.

ILDS attempts to identify the location and amount of a CO_2 leakage at the reservoir level (long before it reaches the surface). By providing such information to the monitoring team at the surface, ample time is provided for proactive intervention rather than reactive responses.

8.8 Enhancement and Evaluation of the ILDS

Development of an intelligent CO_2 ILDS as part of a monitoring system for the CO_2 storage project at Citronelle Dome was explained in detail in the previous sections of this chapter. This system, which was designed based on pattern recognition technology and smart wells, is able to identify the location and amount of CO_2 leakage at the reservoir level (subsurface) using pressure data from PDGs.

A history-matched reservoir simulation model (based on 11 months of actual injection/pressure data) was used for CO_2 leakage modeling. High-frequency real-time pressure streams were processed with a novel technique to form a new data-driven real-time ILDS (RT-ILDS), which was able to detect the leakage characteristics in a short time (less than a day). RT-ILDS also demonstrated high precision in quantifying leakage characteristics subject to complex rate behaviors. The performance of RT-ILDS was examined under different conditions, such as multiple well leakage, availability of an additional monitoring well, uncertainty in the reservoir model, leakage at different vertical locations along the well, and cap-rock leakage. In this section of the chapter, noise behavior in the pressure data and different data cleansing methods are discussed.

8.8.1 RT-ILDS

The ILDS is a data-driven monitoring package that receives real-time pressure data from PDGs and determines the occurrence of a CO_2 leakage, and then calculates the location and amount of leakage. This system was originally designed to receive pressure signals for a time interval comprising one week of hourly signals, or 168 records after the leakage. Summarized pressure data obtained by descriptive statistics was fed into the trained neural networks to find the leakage characteristics. In this system, it was necessary to wait till the end of the time interval to find the leakage characteristics. In this section, a new technique is proposed for the development of the RT-ILDS. In this method the pressure data is analyzed in real time considering the previous trend of the signals. With this method it is possible to determine the leakage characteristics much more quickly (less than a day) than with any other available technique.

In order to process the data and convert it to a format that is appropriate for pattern recognition technology, pressure signals based on 30 different CO_2 leakage scenarios were used. Each scenario corresponded to a simulation run that modeled a specific CO_2 leakage rate in the range 425–2,973 $m^3/$ day with 283 m^3/day increments (15,000–105,000 ft^3/day with 10,000 ft^3/day increments) at one of the three leakage location (wells D-9-6, D-9-7, and D-9-8).

The specifications of the simulation runs and the behavior of the pressure signal for each scenario have been explained in previous sections of this chapter. First, a threshold is assigned as 69 Pa (0.01 psi) for the pressure difference being detected. The pressure difference, Δp, is defined as the difference in pressure behavior with and without a leakage in the system, or ($p_{\text{No leakage}} - p_{\text{Leakage}}$). This threshold is actually equal to the precision of the actual PDG devices currently used in the industry. A PDG with this specific pressure resolution is placed in observation well D-9-8. When this threshold is achieved, data processing starts by considering values of Δp, pressure derivative, Δp average, Δp summation, Δp standard deviation, Δp skewness and kurtosis for the past history of the data. The hourly pressure data for one week for each CO_2 leakage scenario is used to generate the whole data set for the neural network training, calibration, and validation. The first 12 hours of data after the start of the leakage ($\Delta p > 69$ Pa or 0.01 psi) were neglected during the data processing.

8.8.1.1 Neural Network Data Preparation

Development of the RT-ILDS was mainly based on the training, calibration, and validation of neural networks that receive processed real-time pressure data for each CO_2 leakage scenario as the input and the corresponding leakage rate and location as the output. Initially, a neural network was trained to find a pattern between leakage location (output) and the corresponding processed pressure signals. The whole data set for the leakage location neural network consisted of 3,527 data records, which were partitioned into 2,821, 353, and 353 records for training, calibration, and validation, respectively.

The influence of each input parameter on the output (the location of the leakage) was determined by KPI analysis. As shown in Figure 8.29 (left), skewness, standard deviation, and average of Δp have the most impact on the output (location of the leakage). It is worth mentioning that the descriptive

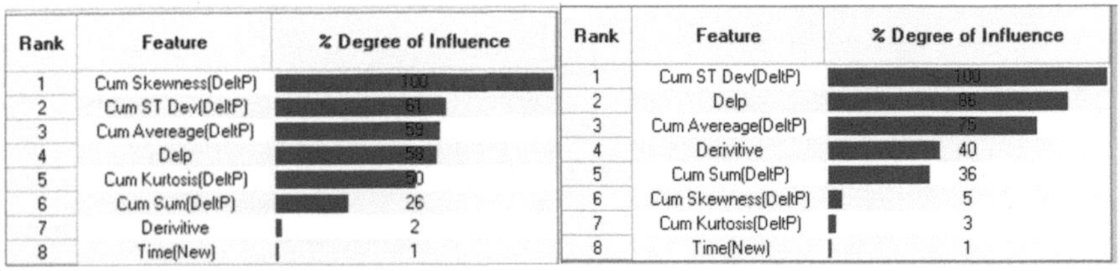

Rank	Feature	% Degree of Influence
1	Cum Skewness(DeltP)	100
2	Cum ST Dev(DeltP)	61
3	Cum Avereage(DeltP)	59
4	Delp	58
5	Cum Kurtosis(DeltP)	30
6	Cum Sum(DeltP)	26
7	Derivitive	2
8	Time(New)	1

Rank	Feature	% Degree of Influence
1	Cum ST Dev(DeltP)	100
2	Delp	88
3	Cum Avereage(DeltP)	75
4	Derivitive	40
5	Cum Sum(DeltP)	36
6	Cum Skewness(DeltP)	5
7	Cum Kurtosis(DeltP)	3
8	Time(New)	1

FIGURE 8.29
Key performance indicator for leakage location (left) and leakage rates at well D-9-8 (right).

statistics for Δp (Figure 8.29) data at each time step is calculated in cumulative basis after a leakage indicator pressure threshold of 69 Pa (0.01 psi) is observed. For example, at time step 24 (after the pressure threshold is detected), average, summation, standard deviation, skewness, and kurtosis were calculated for 24 Δp records (cumulative). Derivative and Δp values are point values at time step 24. The last 12 data records and corresponding calculated parameters will be used in the neural network training.

For determination of the leakage rate, one neural network is trained for each well, separately. The number of data records for each well that is used as input is different due to implementation of the 69 Pa (0.01 psi) threshold as the leakage indicator. For instance, 1,553 records were used to train the leakage rate using the neural network for well D-9-8. Those data records were partitioned into 1,243, 155, and 155 for training, calibration, and validation. The results for the KPI analysis for well D-9-8, showing the impact of the input parameters on the CO₂ leakage rate, are shown in Figure 8.29.

8.8.1.2 Results and Validations

The neural network training process attempts to calculate the most appropriate collection of weights that describes a pattern between the leakage locations and the specified set of input data (pressure signals). The whole process consists of a number of epochs that attempt to minimize the difference or error between the actual and predicted results. It is necessary to calibrate the training process by looking over the training results and finding the best training outcomes. When the error in the calibration data set reaches its minimum value, the training process is completed. The results for all the development processes (training, calibration, and validation) are shown in Figure 8.30 (left: CO₂ leakage location) and Figure 8.30 (CO₂ leakage rate in well D-9-8). For both the leakage location and leakage rates results, R^2 is above 0.99, which represents high precision.

To verify the performance of the RT-ILDS, a set of blind runs (not used for neural network development [training, calibration, and validation]) were designed. As shown in Table 8.7, nine total simulation runs were performed considering the assignment of three CO₂ leakage rates to the possible locations

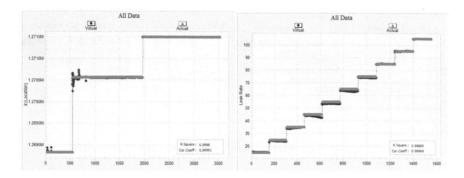

FIGURE 8.30
Neural network results for leakage location (SI conversion: 1 ft = 0.3 m) and neural network results for the leakage rate in well D-9-8 (SI conversion: 1 ft³ = 0.028 m³).

of the leakage (wells D-9-6, D-9-7, and D-9-8). Pressure signals corresponding to each CO_2 leakage scenario were collected, processed by applying the leakage threshold 69 Pa (0.01 psi), and Δp values were generated along with pressure derivative, Δp average, Δp summation, Δp standard deviation, Δp skewness, and Δp kurtosis at each time step. For each leakage scenario, all calculated parameters were used as input to the completed RT-ILDS (in forecast mode) to obtain the prediction for the leakage locations and leakage rate.

All results generated by the RT-ILDS as predictions of the location and rate of leakage for each blind run are shown in Figures 8.31 and 8.32. Figure 8.31 shows the precision of the RT-ILDS in predicting the location of the CO_2 leakage and Figure 8.32 shows the precision of the RT-ILDS in predicting the rate of CO_2 leakage.

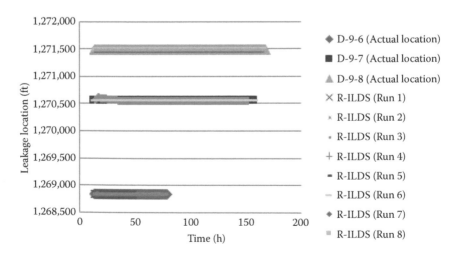

FIGURE 8.31
R-ILDS leakage location prediction for all blind runs (SI conversion: 1 ft = 0.3 m).

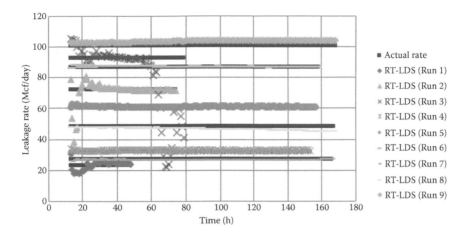

FIGURE 8.32
R-ILDS leakage rate prediction for all blind runs (SI conversion: 1 ft³ = 0.028 m³).

The precision displayed by the RT-ILDS predictions can be quantified by R^2 parameters and the distribution of the errors. The RT-ILDS that was trained for the leakage location has an R^2 almost equal to 1.0. The prediction's error histogram for the locations of wells is shown in Figure 8.33. The average error for the leakage location is about 0.9 m (3 ft) with a maximum error of 14 m (46 ft). The R^2 for the CO_2 leakage rate predictions is 0.998, which indicates a highly precise set of predictions. The percentage error plot for the leak rate at well D-9-8 is shown in Figure 8.34. The maximum error for the leakage rate is less than 9%. The average error for the CO_2 leakage rate predictions is less than 4% at well D-9-8.

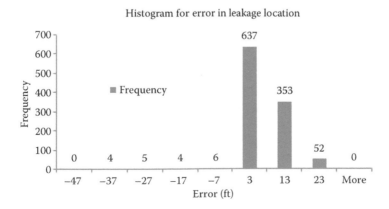

FIGURE 8.33
Histogram for error in the neural network's location prediction (SI conversion: 1 ft = 0.3 m).

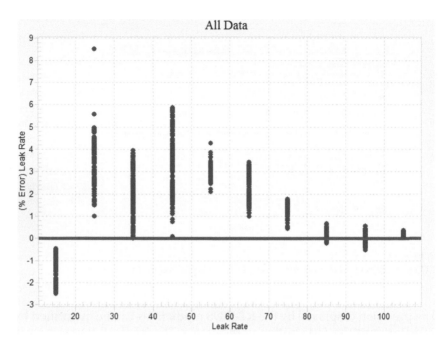

FIGURE 8.34
Neural network prediction errors for leakage rates in well D-9-8 (SI conversion: $1 \text{ ft}^3 = 0.028 \text{ m}^3$).

8.8.2 Detection Time

When CO_2 leakage occurs in the reservoir (from the existing wells, D-9-6, D-9-7, and D-9-8), there is a delay (time lag) until the PDG receives the generated pressure signal. For a given set of reservoir characteristics, the time that it takes to detect the CO_2 leakage depends on the PDG resolution and the amplitude of the pressure signals. The resolution of the PDGs installed in the observation well is 69 Pa (0.01 psi). Therefore, if the amplitude of an induced pressure signal due to the CO_2 leakage is less than the PDG resolution, it would not be possible to detect the leakage.

The other important parameter in leakage detection timing is the amplitude of the pressure signal. The signal amplitude is proportional to the inverse distance of the leakage location to the observation well. The distances of each possible leakage location (wells D-9-6, D-9-7, and D-9-8) to the observation well are shown in Figure 8.35. The induced pressure signal for the cases where each of three wells leaked 1,557 m^3/day (55 Mcf/day) is shown in Figure 8.36.

As the leakage location gets closer to the observation well, the amplitude of the pressure signal increases. RT-ILDS is developed based on the fact that the pressure change threshold of 69 Pa (0.01 psi) can be detected by the PDG. Also, the first 12 pressure data records (after reaching the 69 Pa threshold) are not included in the RT-ILDS development. Based on the mentioned criteria, detection times for different CO_2 leakage rates at each leakage location are

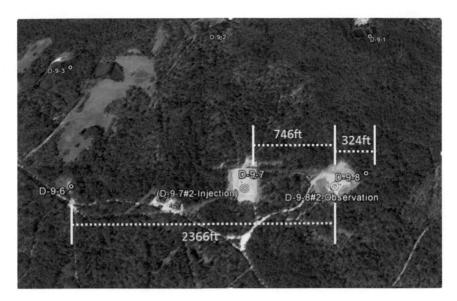

FIGURE 8.35
Distance of possible leakage locations to the observation well (SI conversion: 1 ft = 0.3 m).

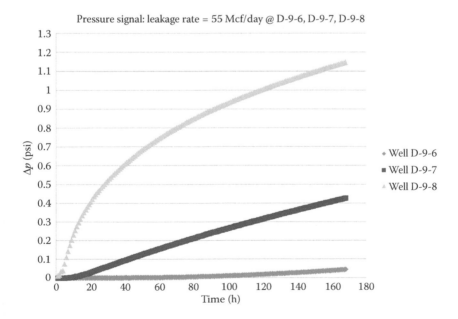

FIGURE 8.36
Comparison of pressure signal amplitude when wells leaked with the same rate (SI conversion: 1 psi = 6.9 kPa, 1 ft^3 = 0.028 m^3).

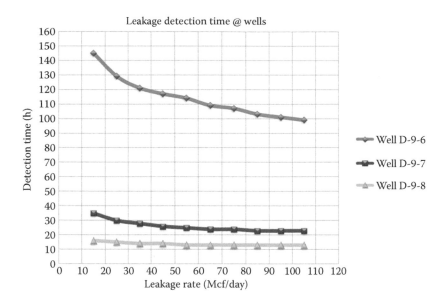

FIGURE 8.37
Detection time for each rate at different locations (SI conversion: 1 ft^3 = 0.028 m^3).

plotted versus CO$_2$ leakage rate in Figure 8.37. As the distance between the source of the leakage and the observation well decreases, the pressure signal amplitudes increase and it takes less time to detect the leakage and provide valid results.

8.8.3 Testing RT-ILDS for Multiple Geological Realizations

The reservoir simulation model for CO$_2$ injection at the Citronelle saline aquifer was developed and history matched with the real field data. This process has been covered in detail in Chapter 5 of this book. The model acknowledges "lateral heterogeneity" in different ways by considering variable sand layer tops, sand thickness, porosity, and permeably values in the reservoir.

Multiple realizations were generated with the aim of changing the characteristics of the fluid flow in this reservoir. This was accomplished by modifying the parameters that control locations along the lateral heterogeneity characteristics. Reservoir porosity, sand layer tops and thickness, as well as vertical to horizontal permeability ratio were the main parameters modified to generate new sets of lateral heterogeneity for different geological realizations that would represent the fluid flow in this reservoir.

All these parameters varied as a function of the original values according to the characteristics displayed in Table 8.8. For each realization, leakage rates equal to 1,982, 1,700, and 1,416 m^3/day (70, 60, and 50 Mcf/day) were assigned to wells D-9-6, D-9-7, and D-9-8, respectively.

TABLE 8.8

Changes in Reservoir Property Parameter

	Variation				
Reservoir Parameters	**2% Up**	**2% Down**	**5% Down**	**10% Up**	**10% Down**
Porosity				x	x
Sand layer top	x	x			
Sand layer thickness			x		
Vertical to horizontal permeability ratio				x	

Note: x is check mark.

The corresponding pressure signals at the observation well were collected, processed, and fed into the RT-ILDS. It should be mentioned that after changing the reservoir characteristics (like porosity or thickness), the initial reservoir pressure and stabilization pressure after the end of the injection process varied (compares with the initial history-matched model). This means that the original $p_{No-leakage}$ would no longer be valid. Therefore, for each given realization a new set of $p_{No-leakage}$ values were needed.

Reservoir pressure signals at the observation well were collected for each realization. At this point, a new Δp was calculated for each realization, with no-leakage pressure data for all cases. As an example, Δp-original and Δp-new for the realization where a porosity of the reservoir lowered by 10% and a CO_2 leakage rate equal to 1,700 m³/day (60 Mcf/day) were assigned to well D-9-7, are shown in Figure 8.38.

Pressure signals from different CO_2 leakage rates and location scenarios and the reservoir characteristic realizations were collected, processed, and

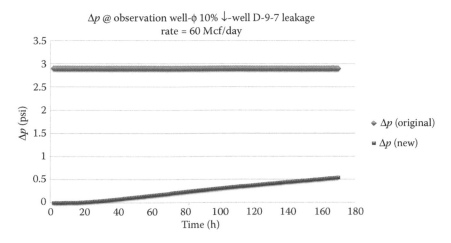

FIGURE 8.38
Original and new Δp at the observation well subject to lowering reservoir porosity (SI conversion: 1 psi = 6.9 kPa).

fed into the RT-ILDS. The impact of the model's specific parameters was studied for the performance of the RT-ILDS. For most cases, changes in the model parameters did not have any significant impact on the RT-ILDS results. The only parameter that considerably impacted the RT-ILDS predictions for both CO_2 leakage rate and location was reservoir porosity. In the reservoir simulation model developed for CO_2 injection at the Citronelle Field, reservoir permeability was calculated by porosity–permeability correlation. Therefore, varying the reservoir porosity directly changed the reservoir permeability. In other words, any changes in reservoir porosity led to a change in the permeability as well. Reservoir permeability plays an important role in fluid flow in the reservoir and consequently affects the pressure signals coming from the observation well. Porosity changes caused different fluid flow behavior and thus different pressure signal behavior. As a result, the RT-ILDS results were impacted by variations in the reservoir porosity.

8.8.4 Detection of Leaks at Different Vertical Locations along the Wells

Based on the reservoir simulation results for the CO_2 distribution and extension in the Citronelle Dome, it was observed that the CO_2 plume reached existing wells in the reservoir mainly in layer 1, as shown in Figure 8.39. All synthetic leakages were thus assigned to the wells at layer 1 (the well was perforated in the model just in that layer). More detailed investigations indicated that the CO_2 plume was in contact with well D-9-7 through nine layers and with well D-9-8 in two layers.

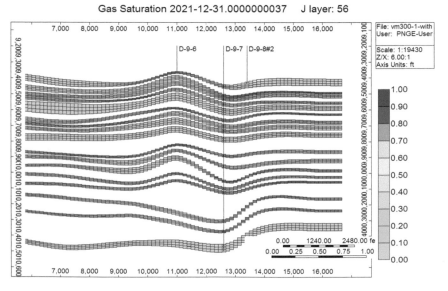

FIGURE 8.39
CO_2 plume extent in different layers (SI conversion: 1 ft = 0.3 m).

This means that the CO_2 leakage could take place at different vertical locations along well D-9-7. For that reason, the changes in the vertical leakage locations were applied to investigate whether the system was capable of detecting the leakage location and leakage rate regardless of where (vertically) within a given well the leak was initiated. It should be mentioned that the two PDGs were installed in well D-9-8 in the first layer of the reservoir.

During the history-matching process, based on the reservoir pressure behavior in the observation well, it was concluded that the transmissibility of the shale layers that were inter-bedded in the sand layers was zero (Figure 8.40). This resulted in alack of any meaningful vertical communications between the sand layers. Therefore, if a leakage took place at well D-9-7 in layer 5, it would not be possible to observe the pressure change at sensors located in layer 1. The pressure change in the PDG located in well D-9-8 when well D-9-7 was leaking from layer 5 at a rate of 1,416 m³/day (50 Mcf/day) is shown in Figure 8.41.

It was assumed that several PDGs were installed at the observation well, exposing the observation well to every sand layer in the reservoir. By making this assumption, it would be possible to measure pressure change due to CO_2 leakage at every layer. Therefore, the corresponding pressure changes (Δp) while wells D-9-7 and D-9-8 were leaking from different vertical locations were recorded, processed, and provided to the RT-ILDS. The RT-ILDS results for the CO_2 leakage locations and rate are shown In Figures 8.42 through 8.44.

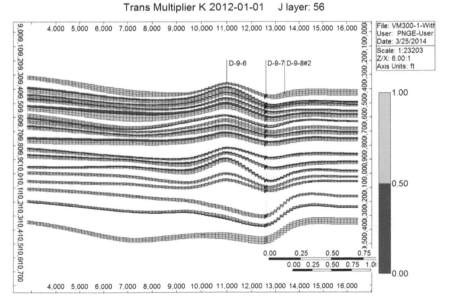

FIGURE 8.40
Transmissibility multiplier for shale layers (SI conversion: 1 ft = 0.3 m).

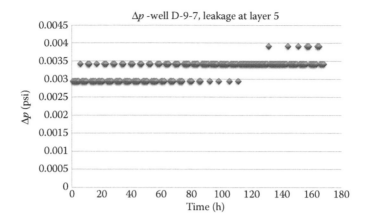

FIGURE 8.41
Pressure change in the observation well when leakage is initiated at layer 5 (SI conversion: 1 psi = 6.9 kPa).

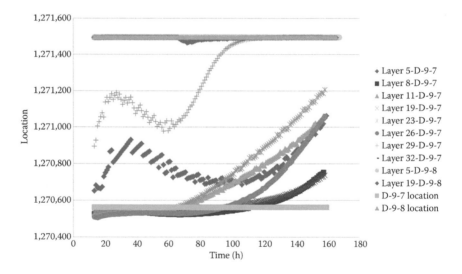

FIGURE 8.42
Leakage location prediction: leakage took place at different vertical locations (SI conversion: 1 ft = 0.3 m).

Based on the results for the leakage location (Figure 8.42), it can be concluded that the RT-ILDS is able to detect the CO$_2$ leakage locations correctly when the CO$_2$ leakage takes place in well D-9-8 at different vertical locations (assuming the existence of PDGs in every layer). When the CO$_2$ leakage takes place at well D-9-7, RT-ILDS predicts the leakage locations correctly, especially up to 80 hours after initiation of the leakage (except the cases where the well leaked from layers 5 and 29). After 80 hours from the detection time, the results started deviating from the actual location of well D-9-7.

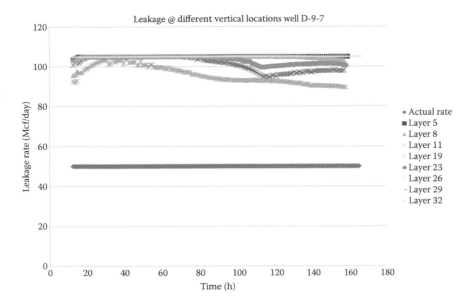

FIGURE 8.43
Leakage rate prediction at well D-9-7 when leakage took place at different vertical locations
(SI conversion: 1 ft³ = 0.028 m³).

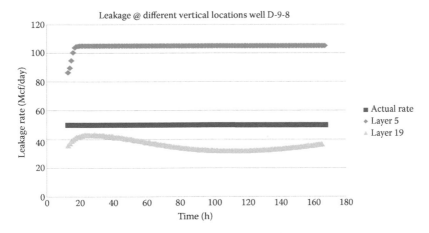

FIGURE 8.44
Leakage rate prediction at well D-9-8 when leakage took place at different vertical locations
(SI conversion: 1 ft³ = 0.028 m³).

A CO_2 leakage rate equal to 1,416 m³/day (50 Mcf/day) was assigned to each
leakage scenario, representing different vertical locations along the well. For
the case where well D-9-7 was leaking (Figure 8.43), the RT-ILDS's leakage rate
predictions were around 2,831 m³/day (100 Mcf/day). When the leakage was
from well D-9-8 (at different layers), RT-ILDS correctly predicted the rate for
the case where CO_2 leakage was from layer 19 (Figure 8.44).

However, the results for the CO$_2$ leakage rate when the leak initiated from layer 5 was not satisfactory. The main reason for not having a correct prediction for the cases where CO$_2$ leakage initiated at different vertical locations is that pressure signals are coming from different layers with completely different reservoir characteristics. Therefore, these pressure signals cannot be exactly the same as the case where the CO$_2$ leakage initiated from layer 1 (the RT-ILDS was developed based on pressure signals for different CO$_2$ leakage scenarios at layer 1).

8.8.5 Impact of Gauge Accuracy or Pressure Drift on RT-ILDS Results

One of the important parameters impacting the accuracy of pressure measurements is pressure sensor drift (PSD). Most of the PDGs experience PSD over their lifetime. PSD can be defined as a gradual malfunction of the sensor that may create offsets in the pressure readings from the original calibrated form (136). Changes in reservoir temperature or pressure make the PDGs respond differently depending on the manufacturing characteristics. The scale of PSD changes according to the working conditions and manufacturing specifications.

PSD can be measured in terms of how much pressure readings deviate from their original values in a year (kPa/year or psi/year). Distributions of different PSD values (137) are shown in Figure 8.45. For RT-ILDS, PSD can act as a CO$_2$ leakage indicator. When $\Delta p = 69$ Pa (0.01 psi) is recorded by the pressure sensor, RT-ILDS reports a leakage and starts processing the data to quantify the leakage characteristics. For example, PSD equal to 6.9 kPa (1 psi/year) generates $\Delta p = 69$ Pa (0.01 psi), almost 88 hours after initiation of the drift.

Based on the different reported values of PSDs (Figure 8.45), the times when the RT-ILDS reports a leakage mistakenly are shown in Figure 8.46. This leakage is due to pressure gauge drift, not an actual induced pressure change.

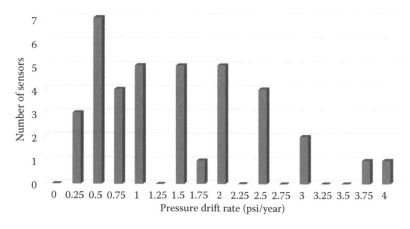

FIGURE 8.45
PSD distribution for the sensors (SI conversion: 1 psi = 6.9 kPa).

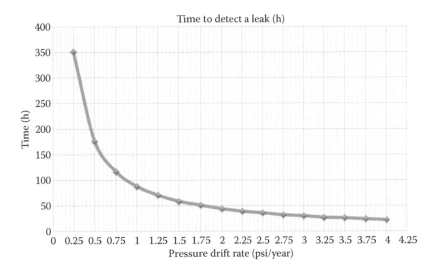

FIGURE 8.46
Time to report a leak based on different PSD values (SI conversion: 1 psi = 6.9 kPa).

As mentioned earlier, PSD can be considered a CO_2 leakage indicator for the RT-ILDS. Therefore, PSD trends (substituting for Δp for actual leakage) over 168 hours were processed and fed to the RT-ILDS. The RT-ILDS prediction results for CO_2 leakage location and rate are shown in Figure 8.47. The RT-ILDS results for leakage locations at early times oscillate between wells D-9-6 and D-9-7. After 80 hours, all the results converge to well D-9-6. This means that PSD makes RT-ILDS report inaccurately that well D-9-6 is leaking.

8.8.6 Use of Well Head Pressure at Injection Well

Typically, there are three different reservoir pressure regimes during injection and post injection. The first period refers to the start of CO_2 injection until it ends (t_1). In this period, the reservoir pressure increases in proportion to the amount of injection and reaches a maximum at the end of injection. When the CO_2 injection ends, there would be a transition time (t_2) when the reservoir pressure decreases until the brine and injected CO_2 reach semi-equilibrium. At the end of the transition time (t_2), the reservoir pressure remains almost constant (or decreases with a slow trend); this can be referred to as a steady-state period (t_3).

These three time cycles are shown in Figure 8.48. The objective of the study presented in this chapter was to develop a RT-ILDS for time cycle t_3, when there is no injection into the field and the reservoir pressure has reached a steady-state trend. During this time period, since CO_2 injection has stopped, there is no fluid flow in the well and the well head pressure will not change (it is possible to have well head pressure during injection, t_1). Therefore, it is not possible to use well head pressure at the injection well for leakage detection in this study.

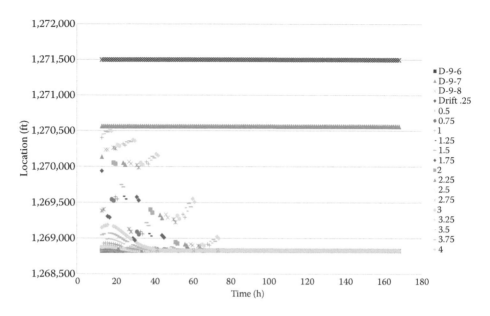

FIGURE 8.47
RT-ILDS prediction for leakage location based on different drift values (SI conversion: 1 ft = 0.3 m).

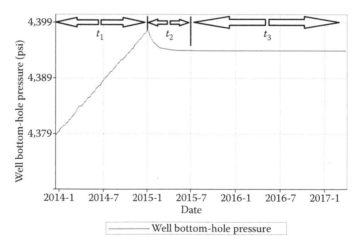

FIGURE 8.48
Different time cycles during and after CO$_2$ injection (SI conversion: 1 psi = 6.9 kPa).

8.8.7 RT-ILDS for Variable CO$_2$ Leakage Rates

In previous sections it was shown that the RT-ILDS is developed by incorporating pressure signals that are generated by step function CO$_2$ leakage rates (Figures 8.49 and 8.50). Step function rates for the CO$_2$ leakages

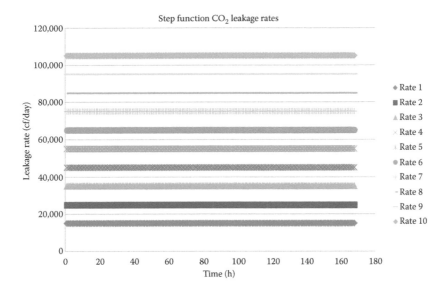

FIGURE 8.49
Step function CO_2 leakage rate (SI conversion: $1 \text{ ft}^3 = 0.028 \text{ m}^3$).

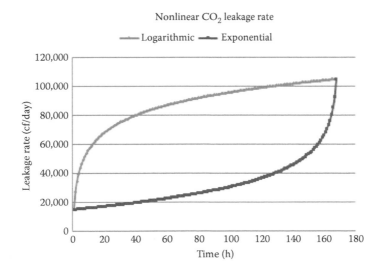

FIGURE 8.50
Logarithmic and exponential CO_2 leakage rates (SI conversion: $1 \text{ ft}^3 = 0.028 \text{ m}^3$).

is defined as leakage initiated at a specific rate that remains constant as time passes. In order to investigate the effect of variable CO_2 leakage rates on the performance of the RT-ILDS, a set of simulation runs were designed and executed to assimilate different CO_2 leakage rate behaviors, such as linear, exponential, and logarithmic.

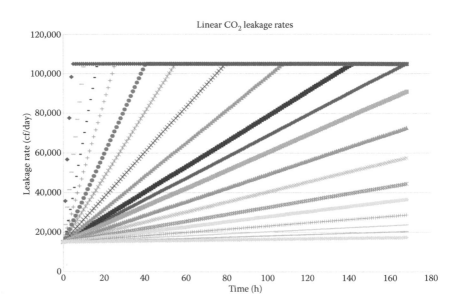

FIGURE 8.51
Linear CO$_2$ leakage rates (SI conversion: 1 ft^3 = 0.028 m^3).

The corresponding pressure signals for each rate function should be included in the development of the leakage detection system. Exponential and logarithmic CO$_2$ leakage rate functions are shown in Figure 8.50. Additionally, 20 different linear CO$_2$ leakage rates were assigned to each possible leakage location (wells D-9-6, D-9-7, and D-9-8) in the reservoir simulation model (60 total simulation runs). Linear CO$_2$ leakage rates are shown in Figure 8.51. The corresponding pressure signals for each CO$_2$ leakage scenario were collected, processed, and sorted to form a data set that is appropriate for pattern recognition technology.

For the detection of CO$_2$ leakage location with different leakage rate functions, all the pressure signals (coming from 60 simulation runs) as function of time and their calculated time-based descriptive statistics were lumped together to form the input data set. Therefore, the input data set included 10,950 data records that were partitioned into training, calibration, and validation sets (80%, 10%, and 10%, respectively). The outputs for this network were three leakage locations (wells D-9-6, D-9-7, and D-9-8).

A back-propagation neural network with 50 neurons in the hidden layer was selected for the training process. The neural network results (virtual versus actual) for the CO$_2$ leakage location are shown in Figure 8.52 (left). The neural network was able to find the pattern between the leakage location and the pressure signals with high precision ($R^2 = 0.99$).

Three neural networks were trained for each well individually to detect the leakage rates. The input data was the same as used for the leakage location training. However, the output is the CO$_2$ leakage rate at each specific time.

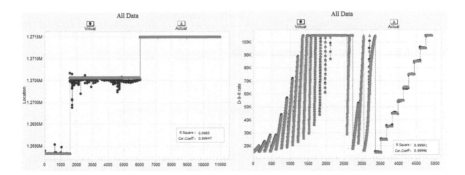

FIGURE 8.52
Left: Neural network results for leakage location (SI conversion: 1 ft = 0.3 m). Right: Neural network results for leakage rates at well D-9-8 (SI conversion: 1 ft^3 = 0.028 m^3).

This is different from the case where the leakage rate remained constant as a function of time.

The neural network architecture was almost the same as previous ones except for the number of neurons in the hidden layers. The results for the CO$_2$ leakage rate at well D-9-8 are shown in Figure 8.52 (right). Neural networks were able to determines a pattern between 32 different CO$_2$ leakage rate functions (as a function of time) and corresponding pressure signals very accurately ($R^2 = 0.99$).

In order to validate the performance of the RT-ILDS, a complex CO$_2$ leakage rate as a function of time was considered for the blind run. This rate function showed logarithmic behavior at the beginning followed by a linear trend. The end part of the rate function presented an exponential characteristic. The rate function for the blind run is shown in Figure 8.53. This rate function was

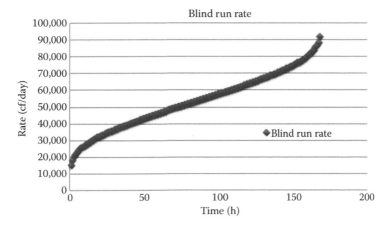

FIGURE 8.53
Rate function for the blind run (SI conversion: 1 ft^3 = 0.028 m^3).

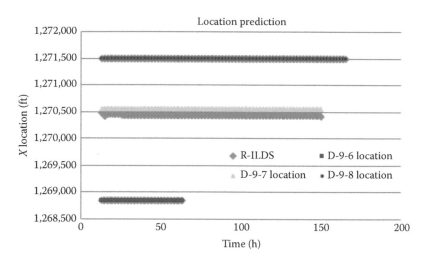

FIGURE 8.54
RT-ILDS prediction for leakage location: variable rate (SI conversion: 1 ft = 0.3 m).

assigned to each of the leakage locations (D-9-6, D-9-7, and D-9-8) as the rate constraint, and the corresponding pressure signal from the observation well (D-9-8) was collected. The pressure signals were processed to obtain real-time Δp and calculate descriptive statistics values to be fed into the RT-ILDS and find the CO_2 leakage location and rate. RT-ILDS predictions for the leakage location and rate (at well D-9-8) are shown in Figures 8.54 and 8.55.

RT-ILDS predictions for the CO_2 leakage locations were reasonably accurate. Additionally, the RT-ILDS was able to predict the location of each well correctly. For the CO_2 leakage rates, in well D-9-8, the RT-ILDS's prediction reproduced the actual rate, especially at early times. The RT-ILDS predicted

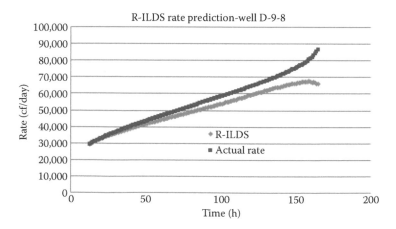

FIGURE 8.55
RT-ILDS prediction for leakage rate in well D-9-8: variable rate (SI conversion: 1 ft³ = 0.028 m³).

Rank	Feature	% Degree of Influence	
1	Cum Sum(DeltP)		100
2	Cum Avereage(DeltP)		80
3	Cum ST Dev(DeltP)		64
4	Cum Skewness(DeltP)		56
5	Derivative		51
6	Delta P		48
7	Time(New)		3
8	Cum Kurtosis(DeltP)		1

FIGURE 8.56
Key performance indicator for CO$_2$ leakage rate in well D-9-8.

just one value for the rate at each time. In order to have a range of rates rather than a single value, "Monte Carlo" simulation was used. Monte Carlo technique or simulation is defined as a statistical method for understanding uncertainty and risk in modeling, forecasting and decision making (138).

A Monte Carlo simulation tends to have the following pattern:

- Identify a range of possible inputs.
- Generate random inputs from a probability distribution over the range.
- Perform a large number of computations with determined inputs.
- Collect, combine, and analyze the results.

The domain of input parameters was defined by carrying out KPI analysis for the leakage rate in well D-9-8 (Figure 8.56).

As shown in in Figure 8.56, cumulative summation (Δp), average (Δp), standard deviation (Δp), and skewness had the most impact on CO$_2$ leakage rate in well D-9-8. To prepare for the Monte Carlo simulation exercise, for each of the mentioned parameters 1,000 random values were generated within "±20%" of the actual value (assuming a rectangular probability distribution). The trained neural network then computed the CO$_2$ leakage rate 1,000 times based on combinations of the generated input variables. Calculated leakage rates were sorted according to their relative frequency and cumulative probability. As an example, at time 162 h after leakage, the actual rate was 2,350 m^3/day (83 Mcf/day), while the RT-ILDS prediction showed 1,909 m^3/day (67.4 Mcf/day). Monte Carlo results provided a leakage rate range (Figure 8.57) that included the actual rate.

8.8.8 Use of the PDG in the Injection Well

Two DPGs were installed in well D-9-8 to transduce real-time pressure data (Figure 8.58). All the studies were performed based on the presence

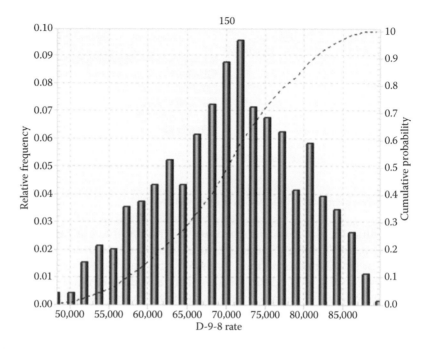

FIGURE 8.57
Relative frequency and cumulative probability for leakage rate (well D-9-8) at time 162 h (SI conversion: 1 ft^3 = 0.028 m^3).

of a PDG at the observation well. In this section, it is assumed that a down-hole pressure gauge is installed in the injection well (D-9-7) rather than the observation well. This may reduce the need for drilling an observation well in the system. All the reservoir simulation runs that addressed 30 different CO$_2$ leakage scenarios (as explained in detail in the previous sections) were used to generate the high-frequency pressure data at the injection well. The

FIGURE 8.58
Location of the injection and observation well in the area of interest.

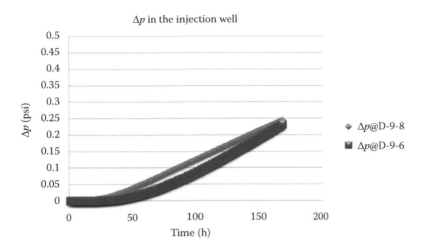

FIGURE 8.59
Pressure signals subject to leakages at wells D-9-6 and D-9-8 (SI conversion: 1 psi = 6.9 kPa).

same procedure was used to develop the RT-ILDS, based on high-frequency pressure data collected at the injection well. The RT-ILDS was developed based on the presence of the PDG at the injection well.

According to the training results, the RT-ILDS was able to predict the CO_2 leakage rates with very good precision (R^2 for the CO_2 leakage rate was more than 0.99 for all three wells, D-9-6, D-9-7, and D-9-8). For the CO_2 leakage location, the RT-ILDS results did not represent the actual locations (the CO_2 leakage location R^2 was 0.49). The reason for not obtaining good results here is that the injection well is located approximately at the same physical distance from the two wells D-9-6 and D-9-8 (Figure 8.58). This symmetric characterization of well locations led to the same pressure signals when wells D-9-6 or D-9-8 were subject to leakage. This can be observed from the pressure signals shown in Figure 8.59.

Since the injection well is located in the middle of the CO_2 plume (based on reservoir characterization), it receives the same pressure signals from different leakages at the same distance from the injection well. Therefore, it is quite difficult to detect the exact location of the leakage correctly. The PDGs should be installed in locations that represent distinct pressure signals from different leakage locations, otherwise the presence of a second monitoring well will be necessary to detect the CO_2 leakage locations correctly.

8.8.9 Leakage from the Cap-Rock

Initially, the reservoir was assumed to have a continuous sealing cap-rock that prevented any communication between the reservoir and the formations above it. After the injection period, pressure on one side of the seal (in the target zone) would increase, leading to a pressure difference across the

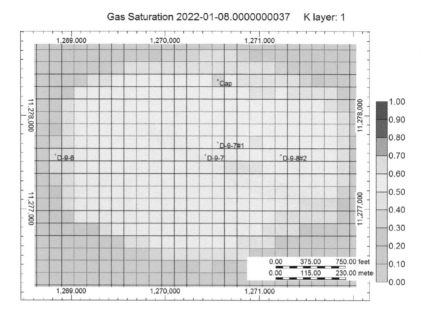

FIGURE 8.60
Cap-rock leakage location (SI conversion: 1 ft = 0.3 m).

cap-rock. As explained previously, when the pressure difference exceeds the fracture pressure, the seal layer may breach and provide a path for the CO$_2$ to migrate to the other layers.

In order to model cap-rock leakage in the reservoir simulator, pressure in the Dantzler sand located on top of the seal (Figure 8.60) was estimated by determining the pressure gradient in the formation and its average depth. This pressure was assigned as the constraint for the cap-rock leakage in the model. The pressure difference between the two layers is the main driving force for CO$_2$ flow through the leakage path. As an example, the reservoir pressure (in the observation well) and the CO$_2$ leakage rate behavior for the case where cap-rock leakage occurs (January 1, 2022) in the north direction of the injection well (Figure 8.60) is shown in Figure 8.61. When the cap-rock fracture is initiated, a large amount of CO$_2$ is released and leaked to the upper layer in a very short time period (less than a day).

This high flow rate of CO$_2$ leakage causes a sharp decline in the reservoir pressure. As the reservoir pressure decreases, the driving force (pressure difference between reservoir and top sand layer) declines and a reduced CO$_2$ leakage rate is observed. Generally, the pressure signal created due to cap-rock leakage has a higher amplitude than well leakage signals (studied in previous sections). Therefore, a different RT-ILDS was developed to be able to detect and quantify the characteristics of the cap-rock leakage.

In order to develop R-ILDS to detect the cap-rock leakage, nine different simulation runs were designed based on the location of the leakage

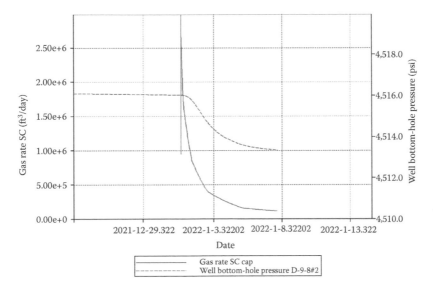

FIGURE 8.61
Pressure behavior in the observation well and CO_2 rate due to cap rock leakage (SI conversion: 1 psi = 6.9 kPa, 1 ft³ = 0.028 m³).

(Figure 8.62). The only constraint for the cap-rock leakage was pressure in the upper layer (Dantzler sand), which was assigned as the bottom-hole pressure for the synthetic well drilled in the leakage location. As mentioned earlier, there is a sharp peak in the CO_2 leakage rate. To eliminate this peak in the CO_2 leakage rate behavior, the cumulative amount of leaked CO_2 was used instead of the instantaneous rate. The training process is exactly the same as already explained in this chapter.

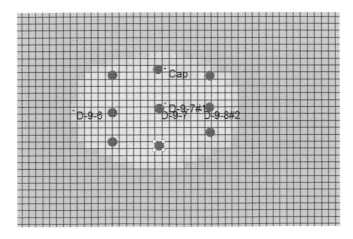

FIGURE 8.62
Nine different locations for cap-rock leakages and three blind runs.

For each leakage scenario, the corresponding pressure signals were processed in real time by descriptive statistics to be used as the input to the neural network. The outputs for the neural network were leakage locations (x and y) and the cumulative amount of leaked CO$_2$. The neural network results for cumulative leaked gas and x coordinate of leakage location were precise, with R^2 equal to 0.97 and 0.99, respectively. For the leakage location y coordinate the neural network predictions were not as accurate as the others (cumulative leaked gas and x coordinate). The reason might be due to the symmetric locations of the cap-rock leakages with respect to the observation well in the y direction.

The final step for validation of the cap-rock RT-ILDS was to design a set of blind runs that were not used during the neural network training process. Three cap-rock leakage locations were considered in the reservoir simulation model (Figure 8.62). Two cap-rock leakage locations (out of three) were inside the range of the locations used for neural network training. For cumulative leaked gas, the RT-ILDS results are almost the same as the actual values for the first two blind run cases, which were located in the range of locations. These results can be seen in Figure 8.63.

For the third blind run, which was located outside the range, the RT-ILDS results overestimated the actual value considerably (Figure 8.64). The x coordinate results were almost the same as the actual locations except for blind run 2. For the y coordinate results, there were noticeable differences between actual values and the RT-ILDS estimations. Overall, the location of the cap-rock leakage can be determined, but it may not be as accurate

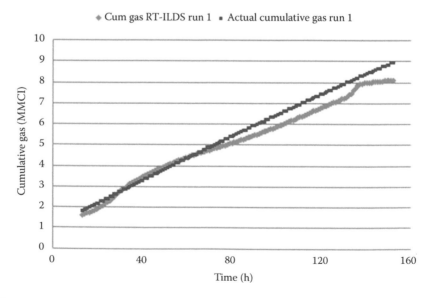

FIGURE 8.63
RT-ILDS prediction for cumulative leaked gas, blind run 1 (SI conversion: 1 ft^3 = 0.028 m^3).

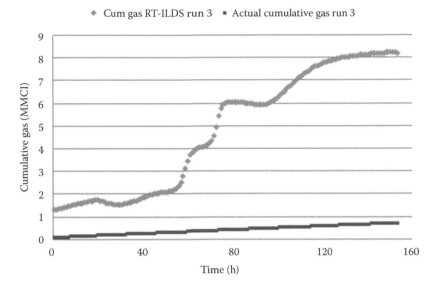

FIGURE 8.64
RT-ILDS prediction for cumulative leaked gas, blind run 3 (SI conversion: 1 ft^3 = 0.028 m^3).

as well-leakage due to the symmetry of the location and the impulsive and uncertain behavior of the leakage.

8.8.10 Multi-Well Leakage

In the previous sections, a single well leakage was studied and analyzed. The remaining question and concern is the possibility of detecting simultaneous leakage from multiple wells. To investigate multi-well leakage, a combination of rates for two and three well leakages was designed and tested. Table 8.9 shows the list of designed scenarios that were examined.

After performing simulation runs (start of leakage at January 1, 2022) based on multi-well leakage scenarios and processing all the corresponding pressure signals, it was required to train a neural network to differentiate between various combinations of well leakages. In this regard, a "leakage index" was defined based on the distance of each well from the observation well. Longer distances from the observation well resulted in the selection of lower values for the leakage index. The index values ranged from 1 to 7 (higher values represent higher pressure signal amplitudes) according to the distance to the observation well and the number of leaking wells.

All the scenarios can be divided into three classes: single-well leakage (indices 1, 2, and 3), two-well leakage (indices 4, 5, and 6), and three-well leakage (index 7). Leakage index values are shown in Table 8.10.

Several neural networks were trained considering different leakage indices as the output and processed pressure signals (Δp) as the input.

TABLE 8.9

CO$_2$ Leakage Rates for Multi-Well Leakage

Two Wells			Three Wells		
Leakage Rate (m^3/day)			Leakage Rate (m^3/day)		
D-9-6	D-9-7	D-9-8	D-9-6	D-9-7	D-9-8
425	425	0	425	425	425
425	1,699	0	425	425	1,699
425	2,974	0	425	425	2,974
1,699	425	0	425	1,699	425
1,699	1,699	0	425	1,699	1,699
1,699	2,974	0	425	1,699	2,974
2,974	425	0	425	2,974	425
2,974	1,699	0	425	2,974	1,699
2,974	2,974	0	425	2,974	2,974
425	0	425	1,699	425	425
425	0	1,699	1,699	425	1,699
425	0	2,974	1,699	425	2,974
1,699	0	425	1,699	1,699	425
1,699	0	1,699	1,699	1,699	1,699
1,699	0	2,974	1,699	1,699	2,974
2,974	0	425	1,699	2,974	425
2,974	0	1,699	1,699	2,974	1,699
2,974	0	2,974	1,699	2,974	2,974
0	425	425	2,974	425	425
0	425	1,699	2,974	425	1,699
0	425	2,974	2,974	425	2,974
0	1,699	425	2,974	1,699	425
0	1,699	1,699	2,974	1,699	1,699
0	1,699	2,974	2,974	1,699	2,974
0	2,974	425	2,974	2,974	425
0	2,974	1,699	2,974	2,974	1,699
0	2,974	2,974	2,974	2,974	2,974

TABLE 8.10

Leakage Index for Different Single- and Multi-Well Leakage Scenarios

Leaking Well	Leakage Index
D-9-6	1
D-9-7	2
D-9-8	3
D-9-6 & D-9-7	4
D-9-6 & D-9-8	5
D-9-7 & D-9-8	6
D-9-6 & D-9-7 & D-9-8	7

It was not possible to obtain reasonable results for the neural networks. The main reason for not achieving acceptable neural network training is that convolution of several pressure signals generated by different combinations of well leakages makes it impossible for networks to catch specific patterns from the final pressure signals. Research to address this issue is currently ongoing at the Petroleum and Natural Gas Engineering Department of West Virginia University.

In order to deconvolve mixed pressure signals (generated by multi-well leakages), the existence of a pressure down-hole gauge was considered in the injection well (in addition to the observation well). The problem was also simplified in a way where only two well leakages were subject to investigation (leakage index values of 4, 5, and 6). Addition of an extra down-hole pressure gauge brought in more information about the pressure signals and the time that signals were observed by the gauges. For this case, a neural network was trained by the generalized regression neural network (GRNN) algorithm. The results for neural network training are shown in Figure 8.65.

By including more pressure gauges in the operation by installing PDGs in the injection well in addition to the observation well, the results for neural network training improved significantly (R^2 equal to 0.9935). As a result, it became possible to differentiate which two wells were leaking by having pressure signals coming from two pressure down-hole gauges. The final

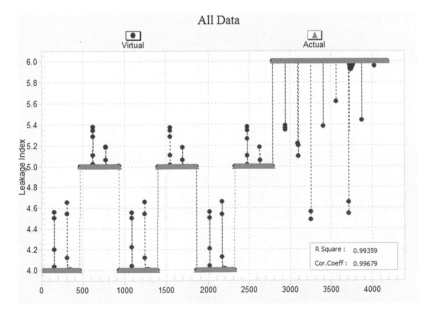

FIGURE 8.65
Neural network results for two-well leakage.

TABLE 8.11

CO_2 Leakage Rates for the Blind Runs: Two-Well Leakages

	Two Wells Leakage Rate (m³/day)		
Run	D-9-6	D-9-7	D-9-8
1	1,132	2,265	0
2	2,265	1,132	0
3	1,132	0	2,265
4	2,265	0	1,132
5	0	1,132	2,265
6	0	2,265	1,132

step was to verify the practicality of the RT-ILDS devolved for the multi-well leakage system. To do this, six new simulation runs considering combinations of two-well leakages were performed. Table 8.10 displays the new (blind) scenarios defined to examine the predictive capabilities of the RT-ILDS in detecting simultaneous leakage from multiple wells (Table 8.11).

The results for the blind runs are shown in Figure 8.66. The RT-ILDS was able to predict the leakage index correctly except for a few hours at early times after the leakages started. Although the probability of two wells leaking simultaneously is reasonably low, with two pressure gauges installed in two different wells (locations) it was possible to say which wells are leaking at the same time.

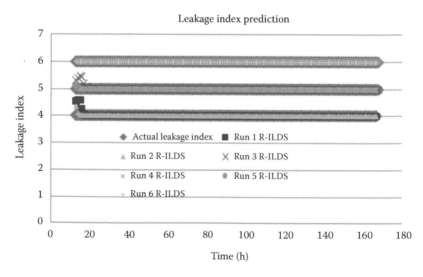

FIGURE 8.66
RT-ILDS predictions for two-well leakages.

8.8.11 Data Cleansing

Interpretation of the PDG data can be challenging due to disturbances like noise and outliers. Noise is a group of data points that scatter around the trend of the overall data but lie in the same neighborhood as the true data. Outliers are data points that lie away from the data trend. Both can be identified from their misalignment with the rest of the data. The real field pressure data gathered from PDGs present some type of noise. In this section, noise associated with pressure data will be analyzed. Additionally, two de-noising methods for cleaning the noisy data are discussed.

8.8.11.1 Determination of Noise Level and Distribution

In the observation well (D-9-8) at Citronelle Field, two PDGs were installed at different depths (2,870 and 2,878 m) in order to provide real-time pressure and temperature readings during and after the injection period. Pressure data is available from August 17, 2012 to November 29, 2013, almost at every minute. It should be mentioned that there are some gaps in the pressure records due to an onsite computer failure. The pressure trends from the PDGs are illustrated in Figure 8.67.

In order to prepare the high-frequency data for pattern recognition (and also the de-noising process) it is necessary to evaluate the noise behavior. There are two main features in the noisy pressure data that need to be analyzed in more detail: noise distribution and noise level. The noise level (139) can be

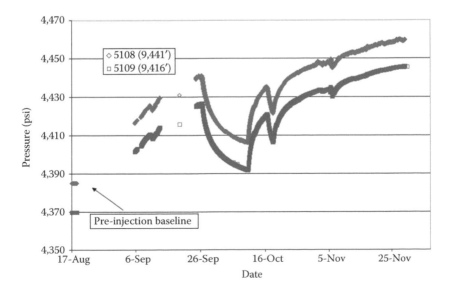

FIGURE 8.67
Monitoring well (D-9-8) PDG data (SI conversion: 1 psi = 6.9 kPa).

determined by knowing the difference between actual data and the fitted curve of the same data over a predefined time interval (with no fluctuation in the data):

$$\text{Noise level} = P_{\text{actual}} - P_{\text{fitted}} \rightarrow \text{Noise Level} = \left(\frac{1}{n-1} \sum_{i=1}^{n} N_i^2 \right)^{1/2}$$

A total of 6,500 pressure records were selected in the interval from September 26 to September 29, 2012 when the pressure trend had no sharp transients. GRNN was used to determine the fitted curve of the selected data. The results of curve fitting are shown in Figure 8.68.

Based on the formula mentioned above, the noise level is 581 Pa (0.08 psi). The maximum and minimum noise levels are equal to 1,248 and −1,944 Pa (0.185 and −0.282 psi), respectively. Also, the frequency distribution of the noises was generated. Based on the results it can be concluded that the noises have a normal or Gaussian distribution (Figure 8.69). Therefore, the noise with the mentioned characteristics will be added to the reservoir simulation pressure scenarios. The noisy and de-noised data would be preprocessed to be transformed into a format that is suitable for pattern recognition analysis.

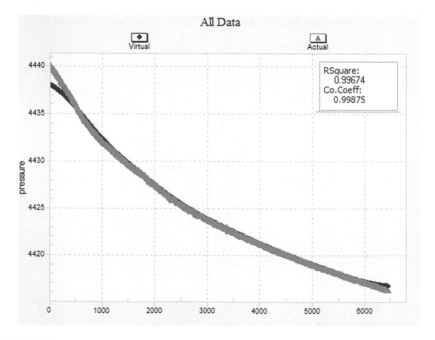

FIGURE 8.68
GRNN results for fitted pressure curve (SI conversion: 1 psi = 6.9 kPa).

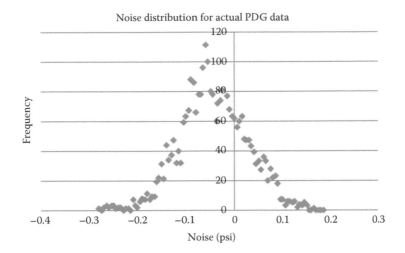

FIGURE 8.69

Noise distribution for actual PDG data: 6,500 records, normal distribution (SI conversion: 1 psi = 6.9 kPa).

8.8.11.2 De-Noising the Pressure Readings

In order to extract the most representative features from the data and reduce fluctuations, a procedure called de-noising is commonly applied. Most de-noising methods tend to smear out sharp features in the data. The method being used in this project is named the "wavelet thresholding" method (140), and generally preserves most of these features.

Wavelet techniques are used to decompose the data to different frequency intervals (components). Each of these components can be investigated separately, resulting in a multi-resolution setting for analyzing the data. Wavelet method advantage over other existing transformation techniques e.g., Fourier transform is its ability to deal with discontinuities (140).

The general de-noising procedure consists of three different steps. Initially, the noisy data should be divided into N levels. After this, for each level from 1 to N, a threshold should be considered and then soft thresholding should be applied to the detail coefficients. Finally, the data is reconstructed using the original approximation coefficients of level N and the modified detail coefficients of levels from 1 to N. The important step in de-noising data is the threshold selection method for each level.

Three threshold selection rules can be implemented: rigorous SURE, heuristics SURE, and fixed form threshold (141). Threshold selection is mainly subjective to the noisy data characteristics. Several methods have also been developed regarding wavelet shrinkage and thresholding. The main two thresholding methods are soft-thresholding and hard-thresholding methods. The main difference between these two methods is that the soft-thresholding method consists of analyzing the difference between the wavelet coefficients and the chosen threshold smoothing the data once the wavelet transform is applied. In the hard-thresholding method, wavelet coefficients whose absolute values are less than the threshold are set to zero. Depending on the scale and particular characteristics of the data, both methods can be used, and the result is cleaned-up data that will still show important details (141).

The pressure data from the reservoir simulation model was considered to be clean with no noise or outlier. For further analysis, it was necessary to add noise with the same characteristic (mentioned earlier) to the pressure data. After adding the noise, data cleansing methods should be applied to the noisy record (generated by adding noise to the clean data). In this study, Daubechies 10 wavelet, in five levels, was used to decompose the noisy data.

After decomposition to five levels, a threshold was assigned to each level to remove the data lying outside the specified level. The processed data from each level was then combined to reconstruct the de-noised data. One example comprising pressure data from the simulator, the same data with normal distributed noise, and cleansed data (de-noised with the wavelet threshold method) when well D-9-6 leaked CO$_2$ at a rate of 850 m^3/day (30 Mcf/day) is shown in Figure 8.70.

The main concern when using the wavelet threshold method for data cleansing is that the de-noised data generally follows the trend of the noisy data. This is mostly the case when leakage rates are low and the corresponding real-time pressure signal changes in the observation well are within a narrow range. This oscillating pattern of de-noised data changes the parameters obtained by data summarization of clean data. In order to alleviate the effect of noise and clean the data in a way that represents the behavior of the pressure trend, GRNN was used. GRNN is a type of probabilistic neural network that just requires a small portion of the data

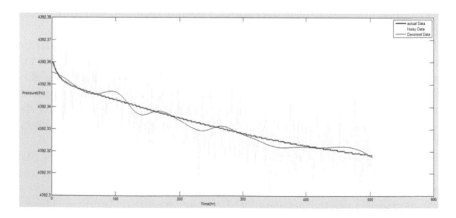

FIGURE 8.70
Pressure data from the simulator (red), with added noise (green) and de-noised (black) when well D-9-6 leaks at a rate of 850 m^3/day (SI conversion: 1 psi = 6.9 kPa).

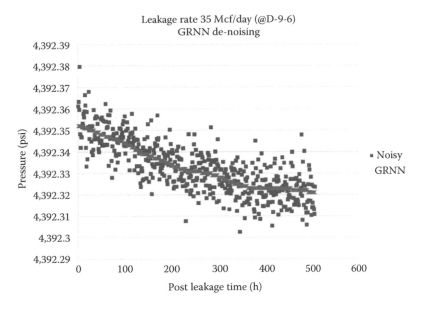

FIGURE 8.71
Noisy and de-noised pressure data using GRNN (SI conversion: 1 psi = 6.9 kPa).

records for training (130). This aspect of GRNN is advantageous as it is able to capture the underlining trend and functionality of the large amount of data with few samples. When high-frequency noisy pressure data need to be processed for leakage detection, it would be better to use GRNN rather than the wavelet threshold method. Because the GRNN uses a smaller portion of data, the presence of noise cannot generally affect the calculated trend. This would be the case especially when the frequency of data increases.

Therefore, GRNN can be considered a very useful tool to de-noise high-frequency pressure data. The results obtained with de-noising pressure records by using GRNN are shown in Figure 8.71. By comparing the resulting trends from the wavelet threshold de-noising method and GRNN with the original clear data, it can be concluded that the GRNN method performs better.

8.8.12 Concluding Remarks

In this chapter a comprehensive study has been presented to develop and then improve ILDS capabilities and test it over different uncertain parameters. A verified and history-matched model was used for CO₂ leakage modeling.

At the beginning, a new data-processing method was proposed to turn ILDS into a fast responsive and real-time detection tool (RT-ILDS). CO₂ leakage characteristics (amount and location) were determined much more quickly that proposed in the previous section. Additionally, minimum detection times for RT-ILDS subject to various leakage locations and rates were determined by considering pressure behavior at the observation well and at the resolution of the PDG. The closer the leaking well to the observation well and the higher the leakage rates, the shorter time was required for leakage detection.

Four different reservoir parameters (porosity, sand layer top, sand layer thickness, and vertical to horizontal permeability ratio) were varied to investigate their effect on the RT-ILDS predictions. The change in reservoir porosity was shown to have a higher impact on RT-ILDS results, especially for CO_2 leakage location and in some cases leakage rate predictions. Sand layer top was the other important parameter that impacted RT-ILDS results.

The ILDS was tested to see if it was possible to detect leakages that took place at different vertical locations along the wells. It was observed that, with the current locations of the PDGs (first sand layer), it would not be possible to sense any pressure changes due to leakage in other layers. Due to the presence of impermeable shale layers between the sands that allowed no inter-communications between layers, PDGs should be installed at each layer specifically to be able to detect leakages at different vertical locations.

The effect of PSD on the performance of the RT-ILDS was studied. For different reported values for PSDs, the time that RT-ILDS reported a leakage (false leakage) and predicted leakage locations were determined. Within the range of pressure drifts of 1.7–27.6 kPa/year (0.25–4 psi/year), it took from 20 to 350 hours for the RT-ILDS to report a leakage. Additionally, the RT-ILDS predicted that leakage took place mostly at well D-9-6.

The use of well-head pressure instead of PDG data was studied in this chapter. Well-head pressure during the injection time (or t_1) would be available. This study's objective was to use pressure data during the stabilization time (or t_3). Therefore, it was not possible to use well-head pressure for leakage detection purposes. Instead, analysis of SCVP was proposed as an alternative solution.

The ability to detect leakages with variable CO_2 leakage rates was added to the RT-ILDS. To do this, multiple CO_2 leakage rates with linear, exponential, and logarithmic behavior were assigned to leakage locations. Corresponding pressure signals were used to train a new neural network. The RT-ILDS gave good results and was tested successfully with a blind run. The RT-ILDS also provided a distribution (by use of Monte Carlo simulation) of predicted CO_2 leakage rates.

The possibility to use the injection well as the monitoring well was investigated as part of this chapter. The procedure of developing an RT-ILDS based on the presence of a PDG in the injection well was almost the same as mentioned at the beginning of this chapter. The injection well RT-ILDS predicted leakage rates with high accuracy but failed to predict the location of leaking wells due to the symmetric locations of the leaking wells with respect to the injection well.

Cap-rock fractures provide a conduit for CO_2 to leak from target formations. Cap-rock leakage behavior (release of a large amount of CO_2 in a very short time) was different from well leakage and was modeled separately in the simulation model considering cap-rock thickness and reservoir pressure in the overlying sand layer. Nine possible locations of cap-rock leakage were proposed within the CO$_2$ plume extent. The RT-ILDS was re-developed and verified based on pressure signals coming from nine simulation and three blind runs. The RT-ILDS predictions were within a reasonable range for cumulative leaked gas and x coordinate of leakage location.

The other concern for R-ILDS was the ability to detect multi-well leakages. A total of 54 simulation runs were performed with two-well and three-well leaking scenarios. With just one observation well it was not possible to distinguish different leakage scenarios. Adding one more observation well in the location of the injection well enabled the RT-ILDS to find which two wells leak simultaneously.

The last part of the chapter studied the behavior of the noise associated with PDG pressure readings. The noise behavior was analyzed based on standard methods (fitted curve analysis). Noise with the same characteristics was added to clean pressure data coming from the reservoir simulation model. Two different de-noising methods – wavelet threshold and GRNN – were implemented to clean the high-frequency noisy data. GRNN showed better de-noising results than the other method.

Bibliography

1. National Climate Data Center. 2017. http://www.ncdc.noaa.gov/indicators/.
2. Bachu, S. 2000. Sequestration of CO_2 in Geological Media: Criteria and Approach for Site Selection in Response to Climate Change. *Energy Conversion and Management*, Vol. 41, pp. 953–970.
3. Kaldi, J. G. and Gibson-Poole, C. M. 2008. *Storage Capacity Estimation, Site Selection and Characterisation for CO_2 Storage Projects.* s.l.: CO2CRC.
4. Cooper, C. 2009. *A Technical Basis for Carbon Dioxide Storage.* UK: CPL Press. ISBN: 978-1-872691-48-0.
5. Carr, N. 2003. IT Doesn't Matter. *Harvard Business Review*, May.
6. McCulloch, W. S. and Pitts, W. 1943. A Logical Calculus of Ideas Immanent in Nervous Activity. *Bulletin of Mathematical Biophysics*, Vol. 5, pp. 115–133.
7. Rosenblatt, F. 1958. The Perceptron: Probabilistic Model for Information Storage and Organization in the Brain. *Psychology Review*, Vol. 65, pp. 386–408.
8. Widrow, B. 1962. Generalization and Information Storage in Networks of Adeline Neurons. *Self-Organizing Systems.* (Eds.) Jacobi, M. C., Goldstein, M. C. and Yovitz, G. D. Chicago, IL: Spartan Books, pp. 435–461.
9. Minsky, M. L. and Papert, S. A. 1969. *Perceptrons.* Cambridge, MA: MIT Press.
10. Hertz, J., Krogh, A. and Palmer, R. G. 1991. *Introduction to the Theory of Neural Computation.* Redwood City, CA: Addison-Wesley.
11. Rumelhart, D. E. and McClelland, J. L. 1986. *Parallel Distributed Processing, Exploration in the Microstructure of Cognition.* Cambridge: MIT Press, Vol. 1: Foundations.
12. Stubbs, D. 1988. Neurocomputers. *M.D. Computing*, Vol. 5, pp. 14–24.
13. Fausett, L. 1994. *Fundamentals of Neural Networks, Architectures, Algorithms, and Applications.* Englewood Cliffs, NJ: Prentice Hall.
14. Barlow, H. B. 1988. Unsupervised Learning. *Neural Computation*, Vol. 1, pp. 295–311.
15. McCord Nelson, M. and Illingworth, W. T. 1990. *A Practical Guide to Neural Nets.* Reading, MA: Addison-Wesley.
16. Haykin, S. 2009. *Neural Networks and Learning Machines*, 3rd edn. Englewood Cliffs, NJ: Prentice Hall.
17. Box, G. E. P. 1976. Science and Statistics. *Journal of the American Statistical Association*, Vol. 71, pp. 791–799.
18. Jolliffe, I. T. 2002. *Principal Component Analysis. Springer Series in Statistics*, 2nd edn. New York, NY: Springer.
19. Shannon, C. E. 1948. A Mathematical Theory of Communication. *Bell Systems Technical Journal*, Vol. 27, pp. 379–423.
20. Mohaghegh, S. D. 2006. Quantifying Uncertainties Associated with Reservoir Simulation Studies Using Surrogate Reservoir Models. *SPE Annual Technical Conference & Exhibition*, September 2006, SPE 102492. San Antonio, TX: Society of Petroleum Engineers (SPE).

21. Mohaghegh, S. D., Hafez, H., Gaskari, R., Haajizadeh, M. and Kenawy, M. 2006. Uncertainty Analysis of a Giant Oil Field in the Middle East Using Surrogate Reservoir Model. *Abu Dhabi International Petroleum Exhibition and Conference— ADIPEC.* SPE 101474, November 2006. Abu Dhabi: SPE.

22. Mohaghegh, S. D., Modavi, A., Hafez, H., Haajizadeh, M., Kenawy, M. and Guruswamy, S. 2006. Development of Surrogate Reservoir Models (SRM) for Fast Track Analysis of Complex Reservoirs. *SPE Intelligent Energy Conference and Exhibition,* April 2006, SPE 99667. Amsterdam: SPE.

23. Mohaghegh, S. D. 2011. Reservoir Simulation and Modeling Based on Pattern Recognition. *SPE Digital Energy Conference and Exhibition.* Woodlands, TX: SPE.

24. Mohaghegh, S. D. et al. 2012. Application of Well-Base Surrogate Reservoir Models (SRMs) to Two Offshore Fields in Saudi Arabia, Case Study. *Western Regional Meeting.* Bakersfield, CA: SPE.

25. CSIRO. 2008. Carbon Storage Trial Set to Go. Science Alert. [Online] December, 2008. http://www.sciencealert.com.au/features/20081303-17040.html.

26. Dance, T., Spencer, L. and Xu, J. 2008. What a Difference a Well Makes. *International Conference on Greenhouse Gas Control Technologies (GHGT-9).* Washington, DC: s.n.

27. Sharma, S. et al. 2009. The CO2CRC Otway Project: Overcoming Challenges from Planning to Execution of Australia's First CCS Project. *Energy Procedia,* Vol. 1, pp. 1965–1972.

28. Szulczewski, M. L. 2013. *The Subsurface Fluid Mechanics of Geologic Carbon Dioxide Storage.* Cambridge, MA: MIT Press.

29. Yamasaki, A. 2003. An Overview of CO_2 Mitigation Options for Global Warming – Emphasizing CO_2 Sequestration Options. *Journal of Chemical Engineering of Japan,* Vol. 36, 361–375.

30. Meer, L. and Egberts, P. 2008. A General Method for Calculating Subsurface CO_2 Storage Capacity. *Offshore Technology Conference,* Houston, TX, 241134. s.n., 2008.

31. Bachu, S. 2008. Comparison between Methodologies Recommended for Estimation of CO_2 Storage Capacity in Geological Media. Carbon Sequestration Leadership Forum.

32. Metz, B. et al. 2005. *Carbon Dioxide Capture and Storage.* New York, NY: Cambridge University Press.

33. Dooleya, J. J. 2004. Accelerated Adoption of Carbon Dioxide Capture and Storage within the United States Utility Industry. *7th International Conference on Greenhouse Gas Control Technologies.* Vancouver, BC, Canada.

34. National Energy Technology Laboratory, NETL. 2010. *Carbon Sequestration Atlas of the United States and Canada (Atlas III).* Morgantown, WV: US Department of Energy.

35. Cooper, C. 2009. *A Technical Basis for Carbon Dioxide Storage.* s.l.: CO_2 Capture Project.

36. Kumar, N. 2008. CO_2 *Sequestration: Understanding the Plume Dynamics and Estimating Risk.* Austin, TX: The University of Texas at Austin.

37. CO2CRC. Cooperative Research Centre for Greenhouse Gas Technologies. [Online] Australian Government's Cooperative Research Centres program. http://www.co2crc.com.au.

38. McGrail, P. 2010. *Metalic Ions in Basalt.* Richland, WA: Pacific Northwest National Laboratory.

39. Initiative, MIT Energy. Carbon Capture & Sequestration Technologies. [Online] Massachusetts Institute of Technology. https://sequestration.mit.edu/tools/projects/sleipner.html.

40. Simmenes, T. et al. 2013. Importance of Pressure Management in CO_2 Storage. *Offshore Technology Conference*, Houston, TX.
41. Riestenberg, D. E. et al. 2009. *CO_2 Sequestration Permitting at the SECARB Mississippi Test Site*. San Antonio, TX: Society of Petroleum Engineers.
42. Koperna, G. J. et al. 2012. The SECARB Anthropogenic Test: The First US Integrated Capture, Transportation, and Storage Test. *Carbon Management Technology Conference*, Orlando, FL.
43. Denbury Resources, Incorporated Plan. 2010. *SECARB Phase III Anthropogenic Test Volume 1 of 2*. Montgomery, AL: Alabama Department of Environmental Management.
44. Haghighat, S. A. et al. 2013. Reservoir Simulation of CO_2 Sequestration in Deep Saline Reservoir, Citronelle Dome, USA. *Twelfth Annual Conference on Carbon Capture, Utilization and Sequestration*, Pittsburgh, PA.
45. Sifuentes, W., Blun, T. M. J. and Giddins, M. A. 2009. *Modeling CO_2 Storage in Aquifers: Assessing the Key Contributors to Uncertainty*. Aberdeen: Offshore Europe.
46. Zhou, Q. et al. 2008. *Sensitivity Study of CO_2 Storage Capacity in Brine Aquifers with Closed Boundaries: Dependence on Hydrogeologic Properties*. s.l.: US Department of Energy.
47. Nghiem, L. et al. 2010. Simulation and Optimization of Trapping Processes for CO_2 Storage in Saline Aquifers. *Journal of Canadian Petroleum Technology*. Vol. 49(8), pp. 15–22. http://dx.doi.org/10.2118/139429-PA.
48. Bennion, D. and Bachu, S. *Drainage and Imbibition Relative Permeability Relationships for Supercritical CO_2/Brine Systems in Intergranular Sandstones, Carbonate, Shale and Anhydrite Rocks*. SPE Europec 2006, Vienna, Austria, SPE 99326.
49. Land, C. S. 1968. Calculation of Imbibition Relative Permeability for Two and Three Phase Flow from Rock Properties. *SPE J*. Vol. 8(2), pp. 149–156, SPE-1942-PA.
50. Harvey, A. 1996. Semiempirical Correlation for Henry's Constants over Large Temperature Ranges. *Am. Inst. Chem. Eng. J*. Vol. 42, pp. 1491–1494.
51. Computer Modelling Group. 2011. *CMG Software Manual*. Houston, TX: Computer Modelling Group.
52. Nelson, C. R. et al. 2005. *Factors Affecting the Potential for CO_2 Leakage from Geologic Sinks*. : NETL
53. Haghighat, S. A. et al. 2013. Pressure History Matching for CO_2 Storage in Saline Aquifers: Case Study for Citronelle Dome. *Carbon Management Technology Conference*, Alexandria.
54. Meckel, T. A. and Hovorka, S. D. 2010. Above-Zone Pressure Monitoring as a Surveillance Tool for Carbon-Sequestration Projects. *SPE International Conference on CO_2 Capture, Storage, and Utilization*, New Orleans, LA.
55. Tao, Q. et al. 2012. Wellbore Leakage Model for Above-Zone Monitoring at Cranfield. *Carbon Management Technology Conference*, Orlando, FL.
56. Torn, A., Torabi, F., Asghari, K. and Mohammadpoor, M. 2012. Effects of Aquifer Parameters on Long-Term Storage of Carbon Dioxide in Saline Aquifers. *Carbon Management Technology Conference*, Orlando, FL.
57. Senel, O. and Nikita Chugunov, N. 2012. CO_2 Injection in a Saline Formation: How Do Additional Data Change Uncertainties in Our Reservoir Simulation Predictions. *Carbon Management Technology Conference*, Orlando, FL.
58. Masoudi, R. et al. 2012. An Integrated Reservoir Simulation-Geomechanical Study on Feasibility of CO_2 Storage in M4 Carbonate Reservoir, Malaysia. *International Petroleum Technology Conference*. Bangkok.

59. Mantilla, C. A. et al. 2009. Inexpensive Assessment of Plume Migration during CO_2 Sequestration. *SPE International Conference on CO_2 Capture, Storage, and Utilization*. San Diego, CA.
60. Krause, M., Perrin, J. C. and Benson, S. M. 2009. Modeling Permeability Distributions in a Sandstone Core for History Matching Coreflood Experiments. *SPE International Conference on CO_2 Capture, Storage*. San Diego, CA.
61. Xiao, C. et al. 2011. Field Testing and Numerical Simulation of Combined CO_2 Enhanced Oil Recovery and Storage in the SACROC Field. *Canadian Unconventional Resources Conference*. Alberta.
62. Mohaghegh, S. D. 2013. *A Critical View of Current State of Reservoir Modeling of Shale Assets*. Pittsburgh, PA: Society of Petroelum Engineers (SPE), 165713.
63. Mohaghegh, S. D. 2017. *Shale Analytics*. Berlin: Springer-Verlag.
64. Kalantari-Dahaghi, A., Mohaghegh, S. and Esmaili, S. 2015. Data-Driven Proxy at Hydraulic Fracture Cluster Level: A Technique for Efficient CO_2-Enhanced Gas Recovery and Storage Assessment in Shale Reservoir. *Journal of Natural Gas Science and Engineering*, Vol. 27.
65. Kalantari-Dahaghi, A., Mohaghegh, S. and Esmaili, S. 2015. Coupling Numerical Simulation and Machine Learning to Model Shale Gas Production at Different Time Resolutions. *Journal of Natural Gas Science and Engineering*, Vol. 25.
66. Kalantari-Dahaghi, A., Esmaili, S. and Mohaghegh, S. 2012. Fast Track Analysis of Shale Numerical Models. *Canadian Unconventional Resources Conference*. SPE 162699. Calgary: Society of Petroleum Engineers (SPE).
67. Mohaghegh, S. and Abdulla, F. 2014. Production Management Decision Analysis Using AI-Based Proxy Modeling of Reservoir Simulations – A Look-Back Case Study. *SPE Annual Technical Conference and Exhibition*. SPE 170664. Amsterdam: Society of Petroleum Engineers.
68. Mohaghegh, S. D. 2017. *Data-Driven Reservoir Modeling*. Richardson, TX: Society of Petroleum Engineers.
69. Intelligent Solutions, Inc. [Online] Intelligent Solutions, Inc., March 7, 2016. [Cited: March 7, 2016.] http://www.intelligentsolutionsinc.com/Technology/TDM.shtml.
70. Jahangiri, H. R. and Zhang, D. 2010. *Optimization of Carbon Dioxide Sequestration and Enhanced Oil Recovery in Oil Reservoir*. Anaheim, CA: s.n., SPE 133594.
71. Gozalpour, F., Ren, S. and Tohidi, B. 2005. CO_2 EOR and Storage in Oil Reservoir. *Oil & Gas Science and Technology*, Vol. 60(3), pp. 537–546.
72. Plasynski, D. and Daminani, S. 2008. *Carbon Sequestration Through Enhanced Oil Recovery*. Albany, OR: US-DOE National Energy Technology Laboratory.
73. U.S. Department of Energy. 2007. *U.S. Department of Energy Carbon Sequestration ATLAS of the United States and Canada*. Morgantown, WV: US DOE National Energy Technology Laboratory.
74. Dicharry, R. M., Perryman, T. L. and Ronquille, J. D. R. 1973. Evaluation and Design of a CO_2 Miscible Flood Project – SACROC Unit, Kelly Snider Field. *Journal of Petroleum Technology*, SPE4083.
75. Raines, M. 2005. Kelly-Snyder (Cisco-Canyon) Fields/SACROC Unit. *West Texas Geological Society: Oil and Gas Fields in West Texas*, Vol. 8, pp. 69–78.
76. Moritis, G. 2003. Kinder Morgan CO_2's Fox: SACROC a "Home Run" for Company. *Oil & Gas Journal*.
77. Vest, E. L. Jr. 1970. Oil Fields of Pennsylvanian–Permian Horseshoe Atoll, West Texas, (Ed.) Halbouty, M. T. *Geology of Giant Petroleum Fields*. Tulsa, OK: American Association of Petroleum Geologists, AAPG Memoir # 14, pp. 185–203.

78. Singh, V. P., Cavanagh, A., Hansen, H., Nazarian, B., Iding, M. and Ringrose, P. S.. 2010. Reservoir Modeling of CO_2 Plume Behavior Calibrated Against Monitoring Data From Sleipner, Norway. Florence, Italy: Society of Petroleum Engineers. ISBN: 978-1-55563-300-4.

79. Walker, D. A., Golonka, J. and Reid, A. M. 1995. *The Effects of Late Paleozic Paleolatitute and Paleogeography on Carbonate Sediment in the Midland Basin, Texas.* (Ed.) Candelaria, M. Midland, TX: West Texas Geological Society Symposium, Publication 91–89, pp. 141–162.

80. Carey, J. W. et al. 2007. Analysis and Performance of Oil Well Cement with 30 Years of CO_2 Exposure from the SACROC Unit. West Texas. *International Journal of Greenhouse Gas Control,* Vol. 8, pp. 75–85.

81. Han, W. S. et al. 2010. Evaluation of Trapping Mechanisms in Geologic CO_2 Sequestration: Case Study of SACROC Northern Platform, A 35-Year CO_2 Injection Site. *American Journal of Science,* Vol. 310, pp. 282–324.

82. Bayat, M. G. et al. 1996. Linking Reservoir Characteristics and Recovery Processes at SACROC – Controlling Wasteful Cycling of Fluids at SACROC While Maximizing Reserves. *Second Annual Subsurface Fluid Control Symposium and Conference.*

83. Helm, L. W. 1959. Carbon Dioxide Solvent Flooding for Increased Oil Recovery. *AIME,* Vol. 216, pp. 225–231.

84. Helm, L. W. and OBrien, L. J. 1971. Carbon Dioxide Test at the Mead-Strawn Field. *Journal of Petroleum Technology,* April, pp. 431–442.

85. Al-Khoury, R. and Bundschuh, J. 2014. *Computational Models for CO2 Geo-sequestration & Compressed Air Energy Storage.* s.l.: CRC Press.

86. Pruess, K. 2006. *Numerical Modeling of CO_2 Sequestration in Geologic Formations – Recent Results and Open Challenges.* Berkeley, CA: Earth Sciences Division, Lawrence Berkeley National Laboratory.

87. Nitao, J. J. 1998. *Reference Manual for the NUFT Flow and Transport Code.* Livermore, CA: Lawrence Livermore National Laboratory, p. 55. UCRL-MA-130651.

88. Hammond, G. E. and Lichtner, P. C. 2010. Field-Scale Model for the Natural Attenuation of Uranium at the Hanford 300 Area Using High-Performance Computing. *Water Resources Research,* Vol. 46.

89. Lichtner, P. C. 2001. *FLOTRAN User's Guide: Two-Phase Nonisothermal Coupled Thermal–Hydrologic–Chemical (THC) Reactive Flow & Transport Code.* Los Alamos, NM: Los Alamos National Laboratory, LA-UR-01–2348.

90. Nghiem, L. et al. 2004. *Modeling CO_2 Storage in Aquifers with a Fully-Coupled Geochemical EOS Compositional Simulator.* Tulsa, OK: Society of Petroleum Engineers.

91. Xu, T. et al. 2010. TOUGHREACT Version 2.0: A Simulator for Subsurface Reactive Transport under Non-Isothermal Multiphase Flow Conditions. *Computers & Geosciences,* Vol. 37, pp. 763–774.

92. Han, W. S. 2008. Evaluation of CO_2 Trapping Mechanisms at the SACROC Northern Platform: Site of 35 Years of CO_2 Injection. *Doctoral dissertation.* Socorro, NM: The New Mexico Institute of Mining and Technology.

93. Lerlertpakdee, P., Jafarpour, B. and Gildon, E. 2014. Efficient Production Optimization with Flow-Network Models. *SPE Journal.*

94. Zhang, Y. and Pau, G. 2012. *Reduced-Order Model Development for CO_2 Storage in Brine Reservoirs.* NRAP Technical Report Series. NRAP-TRS-III-005-2012, Morgantown, WV: U.S. Department of Energy, National Energy Technology Laboratory, p. 20.

95. Zhang, Y. and Sahinidis, N. V. 2013. Uncertainty Quantification in CO_2 Sequestration Using Surrogate Models from Polynomial Chaos Expansion. *Industrial & Engineering Chemistry Research*, Vol. 52, pp. 3121–3132.

96. Van Doren, J. F. M., Markovinovic, R. and Jansen, J. D. 2006. Reduced-Order Optimal Control of Water Flooding Using Proper Orthogonal Decomposition. *Computational Geosciences*, Vol. 10, pp. 137–158.

97. Cardoso, M. A. and Durlofsky, L. J. 2009. Use of Reduced Order Modeling Procedures for Production Optimization. *SPE Journal*.

98. Gildin, E. et al. 2013. Nonlinear Complexity Reduction for Fast Simulation of Flow in Heterogeneous Porous Media. *SPE Reservoir Simulation Symposium*. SPE 163618, Woodlands, TX: SPE.

99. Ifeanyichukwu, P. C., Isehunwa, S. O. and Akpabio, J. U. 2014. A Model for Screening Oil Reservoirs for Carbon Dioxide Flooding. *International Journal of Engineering and Technology*, Vol. 4, 1.

100. Yang, Y. et al. 2009. Reservoir Development Modeling Using Full Physics and Proxy Simulations. *International Petroleum Technology Conference*, Vol. 3, pp. 1449–1461 Doha, Qatar: SPE.

101. Bromhal, G. S. et al. 2014. Evaluation of Rapid Performance Reservoir Models for Quantitative Risk Assessment. *Energy Procedia*. Vol. 114(2017), pp. 5151–5172.

102. He, J. and Durlofsky, L. J. 2014. Reduced-Order Modeling for Compositional Simulation by Use of Trajectory Piecewise Linearization. *SPE Journal*.

103. Hejin, T., Markovinović, R. and Jansen, J. D. 2003. Generation of Low-Order Reservoir Models Using System-Theoretical Concepts. *SPE Reservoir Simulation Symposium*. SPE 79674. Houston, TX: SPE.

104. Klie, H. 2013. Unlocking Fast Reservoir Predictions via Non-Intrusive Reduced Order Models. *SPE Reservoir Simulation Symposium*. SPE 163584. Woodlands, TX: SPE.

105. Chen, H., Klie, H. and Wang, Q. 2013. A Black-Box Stencil Interpolation Method to Accelerate Reservoir Simulations. *SPE Reservoir Simulation Symposium*. SPE 163614. Woodlands, TX: SPE.

106. Fedutenko, E. et al. 2014. Time-Dependent Neural Network Based Proxy Modelling of SAGD Process. *SPE Heavy Oil Conference-Canada*. SPE-170085-MS. Calgary: SPE.

107. Güyagüler, B. et al. 2000. Optimization of Well Placement in a Gulf of Mexico Water Flooding Project. *SPE Annual Technical Conference and Exhibition*. SPE 63221. Dallas, TX: SPE.

108. Artun, E. et al. 2011. Development of Universal Proxy Models for Screening and Optimization of Cyclic Pressure Pulsing in Naturally Fractured Reservoirs. *Journal of Natural Gas Science and Engineering, Artificial Intelligence and Data Mining*, Vol. 3, pp. 667–686.

109. Parada, C. H. and Ertekin, T. 2012. A New Screening Tool for Improved Oil Recovery Methods Using Artificial Neural Networks. *SPE Western Regional Meeting*. SPE 153321. Bakersfield, CA: SPE.

110. Carreras, P. E., Turner, S. E. and Wilkinson, G. T. 2006. Tahiti: Development Strategy Assessment Using Design of Experiments and Response Surface Methods. *SPE Western Regional/AAPG Pacific Section/GSA Cordilleran Section Joint Meeting*. SPE 100656-MS. Alaska, USA: SPE.

111. Li, B. and Friedmann, F. 2005. A Novel Response Surface Methodology Based on "Amplitude Factor" Analysis for Modeling Nonlinear Responses Caused by Both Reservoir and Controllable Factors. *SPE Annual Technical Conference and Exhibition.* SPE 95283. Dallas, TX: SPE.

112. Salhi, M. A. and Van Rijen, M. 2005. Structured Uncertainty Assessment for Fahud Field through the Application of Experimental Design and Response Surface Methods. *SPE Middle East Oil and Gas Show and Conference.* SPE 93529. Kingdom of Bahrain: SPE.

113. Zubarev, D. I. 2009. *Pros and Cons of Applying Proxy-Models as a Substitute for Full Reservoir Simulations.* New Orleans, LA: SPE.

114. Mohaghegh, S. D. 2017. *Data Driven Reservoir Modelling.* Richardson, TX: SPE.

115. Helton, J. C. and Davis, F. J. 2002. *Latin Hypercube Sampling and the Propagation of Uncertainty in Analyses of Complex Systems. Sandia Report.* Albuquerque, NM: Sandia National Laboratories, SAND2001-0417.

116. Hepple, R. P. and Benson, S. M. 2003. Implications of Surface Leakage on the Effectiveness of Geologic Storage of Carbon Dioxide as a Climate Change Mitigation Strategy. *Sixth International Conference on Greenhouse Gas Control Technologies.* Kyoto: s.n.

117. Watson, T. L. and Bachu, S. 2007. Evaluation of the Potential for Gas and CO_2 Leakage along Wellbores. *E&P Environmental and Safety Conference.* Galveston, TX: s.n.

118. Huerta, N. J. 2009. *Studying Fluid Leakage along a Cemented Wellbore.* Austin, TX: The University of Texas at Austin.

119. Tran, D. et al. 2009. Geomechanical Risk Mitigation for CO_2 Sequestration in Saline Aquifers. *SPE Annual Technical Conference and Exhibition.* New Orleans, LA: SPE.

120. Chang, W. C. 2007. *A Simulation Study of Injected CO_2 Migration in the Faulted Reservoir.* Austin, TX: The University of Texas at Austin.

121. Kuuskraa, V. A. 2009. Cost-Effective Remediation Strategies for Storing CO_2 in Geologic Formations. *SPE International Conference on CO_2 Capture, Storage, and Utilization.* San Diego, CA: SPE.

122. Brouwer, D. R. 2004. *Dynamic Water Flood Optimization with Smart Wells Using Optimal Control Theory.* Delft: Deflt University of Technology.

123. Al Omair, A. A. 2007. *Economic Evaluation of Smart Well Technology.* College Station, TX: Texas A&M University.

124. Jansen, J. D. et al. 2009. Closed Loop Reservoir Management. *SPE Reservoir Simulation Symposium.* Woodlands: SPE.

125. Birkholzer, J. T. et al. 2011. Brine Flow Up a Well Caused by Pressure Perturbation from Geologic Carbon Sequestration: Static and Dynamic Evaluations. *International Journal of Greenhouse Gas Control,* Vol. 5, pp. 850–861.

126. Celia, M. A. et al. 2011. Field-Scale Application of a Semi-Analytical Model for Estimation of CO_2 and Brine Leakage along Old Wells. *International Journal of Greenhouse Gas Control,* Vol. 5, pp. 257–269.

127. Jung, Y., Quanlin, Z. and Birkholzer, J. T. 2011. Early Detection of Brine and CO_2 Leakage through Abandoned Wells Using Pressure and Surface-Deformation Monitoring Data: Concept and Demonstration. *Advances in Water Resources,* Vol. 62, Part C, pp. 555–569.

128. Zeidouni, M. and Pooladi-Darvish, M. 2010. Characterization of Leakage through Cap-Rock with Application to CO_2 Storage in Aquifers – Single Injector and Single Monitoring Well. *Canadian Unconventional Resources and International Petroleum Conference*. Calgary: s.n.

129. Sun, A. Y. and Nicot, J. P. 2013. Inversion of Pressure Anomaly Data for Detecting Leakage at Geologic Carbon Sequestration Sites. *Advances in Water Resources*, Vol. 56, pp. 49–60.

130. Jalali, J. 2010. *Artificial Neural Networks for Reservoir Level Detection of CO_2 Seepage Location Using Permanent Down-Hole Pressure Data*. Morgantown, WV: West Virginia University.

131. Haghighat, S. A. et al. 2013. Using Big Data and Smart Field Technology for Detecting Leakage in a CO_2 Storage Project. *SPE Annual Technical Conference and Exhibition*. New Orleans, LA: SPE.

132. Wilson, T. and Wells, A. W. 2010. Multi-Frequency EM Surveys Help Identify Possible Near-Surface Migration Pathways in Areas Surrounding a CO_2 Injection Well: San Juan Basin, New Mexico, USA. 3, *Environmental and Engineering Geophysical Society, Fast Times*, Vol. 15, pp. 43–53.

133. Lozzio, M. et al. 2010. *Quantifying the Risk of CO_2 Leakage through Wellbores*. New Orleans, LA: SPE.

134. Wikipedia. Wikipedia, the free encyclopedia. [Online] http://en.wikipedia.org/wiki/Descriptive_statistics.

135. Intelligent Solutions, Inc. 2014. *IDEA User Manual*. Morgantown, WV: Intelligent Solutions, Inc.

136. Solinst. Solinst Canada Ltd. [Online] Solinst Canada Ltd. http://www.solinst.com/products/dataloggers-and-telemetry/3001-levelogger-series/technical-bulletins/understanding-pressure-sensor-drift.php.

137. Reiter, J., Murphy, D. and Larson, N. 2012. Drift Measurements in Pressure Sensors. *Ocean Sciences Meeting*. Salt Lake City, UT: s.n.

138. Wikipedia. Wikipedia. [Online] Wikipedia, the free encyclopedia. http://en.wikipedia.org/wiki/Monte_Carlo_method.

139. Kin, K. C. 2001. *Permanent Down-Hole Gauge Data Interpretation*. Stanford, CA: Stanford University.

140. Graps, A. 1995. An Introduction to Wavelets. *IEEE Computational Science & Engineering*, Vol. 2, pp. 50–61.

141. Mathworks. 1994. Mathworks. [Online] MathWorks, http://www.mathworks.com/.

142. Kuhn, T. S. 1996. *The Structure of Scientific Revolutions*. Chicago, IL: University of Chicago Press, 0-226-45807-5.

143. Kuhn, T. S. 1984. *The Essential Tension: Selected Studies in Scientific Tradition and Change*. Chicago, IL: University of Chicago Press, 0-226-45806-7.

144. Bell, G., Hey, T. and Szalary, A. 2009. Beyond the Data Deluge. *Science*, Vol. 323, pp. 1297–1298.

145. Freeman, E. 1983. *The Relevance of Charles Peirce*. La Sall, IL: Monist Library of Philosophy, pp. 157–158.

146. Lukasiewicz, J. 1963. *Elements of Nathematical Logic [Original Title: Elementy logiki matematycznej]*. New York, NY: MacMillan.

147. Black, M. 1937. Vagueness: An Exercise in Logical Analysis. *Philosophy of Science*, Vol. 4, pp. 427–455.

148. Zadeh, L. A. 1965. *Fuzzy Sets. Information and Control*, Vol. 8, pp. 338–353.

149. Kosko, B. 1991. *Fuzzy Thinking.* New York, NY: Hyperion.
150. McNeill, D. and Freiberger, P. 1993. *Fuzzy Logic.* New York, NY: Simon & Schuster.
151. Eberhart, R., Simpson, P. and Dobbins, R. 1996. *Computational Intelligence PC Tools.* Orlando, FL: Academic Press.
152. Ross, T. 1995. *Fuzzy Logic With Engineering Applications.* New York, NY: McGraw-Hill.
153. Jamshidi, M. et al. (Eds.) 1993. *Fuzzy Logic and Control: Software and Hardware Applications.* Englewood Cliffs, NJ: Prentice Hall.
154. Bbauska, R. 2009. *Fuzzy and Neural Control.* Delft Center for Systems and Control. Delft, Holland.
155. Pelham Box, G. E. 1976. *Science and Statistics.* p. 792.
156. Thakur, G. C. 1996. What Is Reservoir Management? *Journal of Petroleum Technology,* Vol. 48, pp. 520–525.
157. Corporation, Chevron. 2012. Reservoir Management. *Chevron Corporation Website.* [Online] Chevron Corporation Website. http://www.chevron.com/deliveringenergy/oil/reservoirmanagement/.
158. Mata, D., Gaskari, R. and Mohaghegh, S. D. October 2007. *Field-Wide Reservoir Characterization Based on a New Technique of Production Data Analysis.* Lexington, KY: SPE, SPE 111205.
159. Gomez, Y., Khazaeni, Y., Mohaghegh, S. D. and Gaskari, R. 2009. *Top-Down Intelligent Reservoir Modeling (TDIRM).* New Orleans, LA: SPE, SPE 124204.
160. Gaskari, R., Mohaghegh, S.D. and Jalali, J. November 2007. An Integrated Technique for Production Data Analysis (PDA) with Application to Mature Fields. *SPE Production & Operations Journal,* Vol. 22, pp. 403–416.
161. Mohaghegh, S. D. and Gaskari, R. 2009. An Intelligent System's Approach for Revitalization of Brown Fields Using Only Production Rate Data. *International Journal of Engineering,* Vol. 22, pp. 89–106.
162. Kalantari, A. M., Mohaghegh, S. D. and Khazaeni, Y. 2010. New Insight into Integrated Reservoir Management Using Top-Down, Intelligent Reservoir Modeling Technique; Application to a Giant and Complex Oil Field in the Middle East. *SPE Western Regional Conference & Exhibition.* Anaheim, CA: SPE, SPE 132621.
163. Khazaeni, Y. and Mohaghegh, S. D. 2011. Intelligent Production Modeling Using Full Field Pattern Recognition. *SPE Reservoir Evaluation and Engineering Journal,* Vol. 14, pp. 735–749.
164. Mohaghegh, S. D. 2011. Reservoir Simulation and Modeling Based on Artificial Intelligence and Data Mining (AI&DM). *Journal of Natural Gas Science and Engineering,* Vol. 3, pp. 697–705.
165. Maysami, M., Gaskari, R. and Mohaghegh, S. D. 2013. Data Driven Analytics in Powder River Basin, WY. *SPE Annual Technical Conference and Exhibition.* New Orleans, LA: SPE, SPE 166111.
166. Zargari, S. and Mohaghegh, S. D. 2010. Development Strategies for Bakken Shale Formation. *SPE Eastern Regional Conference & Exhibition.* Morgantown, WV: SPE, SPE 139032.
167. Esmaili, S., Kalantari, M. and Mohaghegh, S. 2012. Modeling and History Matching Hydrocarbon Production from Marcellus Shale Using Data Mining and Pattern Recognition Technologies. *SPE Eastern Regional Conference.* Lexington, KY: SPE, SPE 161184.

168. Grujic, O., Mohaghegh, S. D. and Bromhal, G. 2010. Fast Track Reservoir Modeling of Shale Formations in the Appalachian Basin. Application to Lower Huron Shale in Eastern Kentucky. *SPE Eastern Regional Conference & Exhibition.* Morgantown, WV: SPE, SPE 139101.

169. Kalantari, A. M. and Mohaghegh, S. D. 2011. A New Practical Approach in Modeling and Simulation of Shale Gas Reservoirs: Application to New Albany Shale. *International Journal of Oil, Gas and Coal Technology,* Vol. 4, pp. 104–133.

170. Mohaghegh, S. D., Grujic, O., Zargari, S., Kalantari, A. M. and Bromhal, G. 2012. Top-Down, Intelligent Reservoir Modeling of Oil and Gas Producing Shale Reservoirs; Case Studies. *International Journal of Oil, Gas and Coal Technology,* Vol. 5, pp. 3–28.

171. Haghighat, A., Mohaghegh, S. D., Gholami, V. and Moreno, D. 2014. Production Analysis of a Niobrara Field Using Intelligent Top-Down Modeling. *SPE Western North American and Rocky Mountain Joint Regional Meeting.* Denver, CO: SPE, SPE 169573.

172. Aurenhammer, F. 1991. Voronoi Diagrams – A Survey of a Fundamental Geometric Data Structure. *ACM Computing Surveys,* Vol. 23, pp. 345–405.

173. Sayarpour, M., Kabir, C. S. and Lake, L. W. 2009. Field Applications of Capacitance-Resistance Models in Waterfloods. *SPE Reservoir Evaluation & Engineering.* Vol. 12(6), pp. 853–864. doi: 10.2118/114983-PA.

174. Franklin, S. and Graessers, A. 1997. Is It an Agent, or Just a Program?: A Taxonomy for Autonomous Agents. (Eds.) Wooldridge, M. J., Jennings, N. and Müller, J. P. *Intelligent Agents III Agent Theories, Architectures, and Languages.* Berlin: Springer, pp. 21–35.

175. Stufflebeam, R. Introduction to Intelligent Agents. *Consortium on Cognitive Science Instruction.* [Online] The Mind Project. [Cited: June 23, 2015.] http://www.mind.ilstu.edu/curriculum/ants_nasa/intelligent_agents.php.

176. Swan, A. R. H. and Sandilands, M. 1995. *Introduction to Geological Data Analysis.* Oxford: Blackwell.

177. Jensen, J. L., Lake, L. W., Corbett, P. W. M. and Goggin, D. J. 2000. *Statistics for Petroleum Engineers and Geoscientists,* 2nd edn. Amsterdam: Elsevier.

178. Davis, J. C. 2002. *Statistics and Data Analysis in Geology,* 3rd edn. New York, NY: Wiley.

179. King, A. D.-G. and Michael, J. 2007. *Streamline Simulation: Theory and Practice.* Richardson, TX: SPE.

180. Jalali, J., Mohaghegh, S. D. and Gaskari, R. 2006. Identifying Infill Locations and Underperformer Wells in Mature Fields Using Monthly Production Rate Data, Carthage Field, Cotton Valley Formation. *SPE Eastern Regional Meeting,* Canton, OH, SPE 104550.

181. Gaskari, R. and Mohaghegh, S. D. 2006. Estimating Major and Minor Natural Fracture Pattern in Gas Shales Using Production Data. *SPE Eastern Regional Meeting,* Canton, OH, SPE 104554.

182. Mohaghegh, S. D. 2009. *Top-Down Intelligent Reservoir Modeling (TDIRM); A New Approach in Reservoir Modeling by Integrating Classis Reservoir Engineering with Artificial Intelligence and Data Mining Techniques.* Denver, CO: American Association of Petroleum Geologists (AAPG).

183. Mohaghegh, S. D. and Bromhal, G. 2010. *Top-Down Modeling; Practical, Fast Track, Reservoir Simulation & Modeling for Shale Formations.* Austin, TX: AAPG/SEG/SPE/SPWLA Hedberg Conference.

184. Esmaili, S. and Mohaghegh, S. D. 2015. Full Field Reservoir Modeling of Shale Assets Using Advanced Data-Driven Analytics. *Geoscience Frontiers*, Vol. 7, http://dx.doi.org/10.1016/j.gsf.2014.12.006. http://www.sciencedirect.com/science/article/pii/S1674987114001649#

185. Mohaghegh, S. D., Gaskari, R., Maysami, M. and Khazaeni, Y. 2014. *Data-Driven Reservoir Management of a Giant Mature Oilfield in the Middle East*. Amsterdam: SPE, SPE 170660.

186. Al-Sharhan, A. S. 1993. Bu Hasa Field – United Arab Emirates, Rub' al Khali Basin, Abu Dhabi. *Structural Traps VIII. Treaties of Petroleum Geology, Atlas of Oil and Gas Fields*. (Eds.) Foster, N. H. and Beaumont, E. A. Washington, DC: AAPG, pp. 69–97.

187. Dickerson, M. T., Goodrich, M. T., Dickerson, M. D. and Zhuo, Y. D. 2011. Round-Trip Voronoi Diagrams and Doubling Density in Geographic Networks. *Transactions on Computational Science*, Vol. 14, pp. 211–238.

188. Höppner, F. et al. 1997. *Fuzzy Cluster Analysis: Methods for Classification, Data Analysis and Image Recognition*. New York, NY: Wiley.

189. Mohaghegh, S. D. 2013. Reservoir Modeling of Shale Formations. *Journal of Natural Gas Science and Engineering*, Vol. 12, pp. 22–33.

190. Hunt, G. How the US Shale Boom Will Change the World. OilPrice.com. [Online] February 15, 2012. [Cited: November 30, 2015.] http://oilprice.com/Energy/Natural-Gas/How-the-US-Shale-Boom-Will-Change-the-World.html.

191. Genesis. 2011. The North Texas Barnett Shale Opens New Energy Era, North American Shale Revolution. Energyintel.com. [Online] Sept. – Dec. www.energyintel.com.

192. Dershowitz, W. and Doe, T. W. 2011. *Modeling Complexities of Natural Fracturing Key in Gas Shales*. *American Oil and Gas Reporter*, August.

193. Ilk, D., Rushing, J. A., Perego, A. D. and Blasingame, T. A. 2008. *Exponential vs. Hyperbolic Decline in Tight Gas Sands: Understanding the Origin and Implications for Reserve Estimates Using Arp's; Decline Curves*. s.l.: SPE.

194. Valko, P. P. and Lee, W. J. 2010. *A Better Way to Forecast Production from Unconventional Gas Wells*. s.l.: SPE.

195. Duong, A. N. 2010. *An Unconventional Rate Decline Approach for Tight and Fracture-Dominated Gas Wells*. s.l.: SPE.

196. Mohaghegh, S. D. 2015. Formation vs. Completion: Determining the Main Drivers behind Production from Shale? A Case Study Using Data-Driven Analytics. *Unconventional Resources Technology Conference*. San Antonio, TX: SPE, July, URTeC 2147904.

197. Cipolla, C. L. 2011. Stimulated Volume and Fracture Structure, the Keys to Shale-Gas Well Performance? Amelia Island Plantation. *SPEE 49th Annual Meeting*, Amelia Island Plantation, Florida.

198. Inamdar, A., Ogundare, T., Purcell, D., Malpani, R., Atwood, K., Brook, K. and Erwemi, A. 2011. *Pilot Test Stimulation Approach*. American Oil and Gas Reporter, June, 61–67.

199. Ciezobka, J. *Marcellus Shale Gas Project*. s.l.: RPSEA, Annual Report, February 2012.

200. Shannon, C. E. 1948. A Mathematical Theory of Communication. *The Bell System Technical Journal*, Vol. 27, pp. 379–423, 623–656.

201. Bezdek, J. 1984. The Fuzzy c-Means Clustering Algorithm. *Computer and Geosciences*, Vol. 10(2–3), pp. 191–203.

202. Coulter, G. R. and Menzie, D. E. 1973. The Design of Re-frac Treatments for Restimulation of Subsurface Formations. *Rocky Mountain Regional Meeting,* Casper, WY: SPE, SPE 4400.

203. Mohaghegh, S., Hefner, H. and Ameri, S. 1996. Fracture Optimization eXpert (FOX): How Computational Intelligence Helps the Bottom-Line in Gas Storage. *SPE Eastern Regional Conference and Exhibition.* Columbus, OH: SPE, SPE 37341.

204. Mohaghegh, S., Balan, B., McVey, D. and Ameri, S. 1996. A Hybrid Neuro-Genetic Approach to Hydraulic Fracture Treatment Design and Optimization. *SPE Annual Technical Conference & Exhibition (ATCE).* Denver, CO: SPE, SPE 36602.

205. Mohaghegh, S., Platon, V. and Ameri, S. 1998. Candidate Selection for Stimulation of Gas Storage Wells Using Available Data with Neural Networks and Genetic Algorithms. *SPE Eastern Regional Conference and Exhibition.* Pittsburgh, PA: SPE, SPE 51080.

206. Mohaghegh, S., Mohamad, K., Popa, A. S. and Ameri, S. 1999. Performance Drivers in Restimulation of Gas Storage Wells. *SPE Eastern Regional Conference and Exhibition.* Charleston, WV: SPE, SPE 57453.

207. Reeves, S. R., Hill, D. G., Tiner, R. L., Bastian, P. A., Conway, M. W. and Mohaghegh, S. D. 1999. Restimulation of Tight Gas Sand Wells in the Rocky Mountain Region. *SPE Rocky Mountain Region Meeting.* Gillette, WY: SPE, SPE 55627.

208. Reeves, S. R., Hill, D. G., Hopkins, C. W., Conway, M. W., Tiner, R. L. and Mohaghegh, S. D. 1999. Restimulation Technology for Tight Gas Sand Wells. *SPE Technical Conference and Exhibition (ATCE).* Houston, TX: SPE, SPE 56482.

209. Mohaghegh, S., Reeves, S. and Hill, D. 2000. Development of an Intelligent Systems Approach to Restimulation Candidate Selection. *SPE Gas Technology Symposium.* Calgary: SPE, SPE 59767.

210. Reeves, S., Bastian, P., Spivey, J., Flumerfelt, R., Mohaghegh, S. and Koperna, G. 2000. Benchmarking of Restimulation Candidate Selection Techniques in Layered, Tight Gas Sand Formations Using Reservoir Simulation. *SPE Annual Technical Conference and Exhibition (ATCE).* Dallas, TX: SPE, SPE 63096.

211. Mohaghegh, S., Gaskari, R., Popa, A., Ameri, S. and Wolhart, S. 2001. Identifying Best Practices in Hydraulic Fracturing Using Virtual Intelligence Techniques. *SPE Eastern Regional Conference and Exhibition.* North Canton, OH: SPE, SPE 72385.

212. Mohaghegh, S., Popa, A., Gaskari, R., Ameri, S. and Wolhart, S. 2002. Identifying Successful Practices in Hydraulic Fracturing Using Intelligence Data Mining Tools; Application to the Codell Formation in the DJ Basin. *SPE Annual Conference and Exhibition (ATCE).* San Antonio, TX: SPE, SPE 77597.

213. Mohaghegh, S. D. 2003. Essential Components of an Integrated Data Mining Tool for the Oil & Gas Industry, with an Example Application in the DJ Basin. *SPE Annual Conference and Exhibition (ATCE).* Denver, CO: SPE, SPE 84441.

214. Mohaghegh, S. D., Gaskari, R., Popa, A., Salehi, I. and Ameri, S. 2005. Analysis of Best Hydraulic Fracturing Practices in the Golden Trend Fields of Oklahoma. *SPE Annual Conference and Exhibition (ATCE).* Dallas, TX: SPE, SPE 95942.

215. Mohaghegh, S. D. and Gaskari, R. 2005. A Soft Computing-Based Method for the Identification of Best Practices, with Application in Petroleum Industry. *IEEE International Conference on Computational Intelligence for Measurement Systems& Applications.* Taormina, Sicily: IEEE.

216. Mayr, E. 1988. *Toward a new Philosophy of Biology: Observations of an Evolutionist.* Cambridge, MA: Belknap Press.

217. Koza, J. R. 1992. *Genetic Programming, On the Programming of Computers by Means of Natural Selection.* Cambridge, MA: MIT Press.
218. Fogel, D. B. 1995. *Evolutionary Computation, Toward a New Philosophy of Machine Intelligence.* Piscataway, NJ: IEEE Press.
219. Michalewicz, Z. 1992. *Genetic Algorithms + Data Structure = Evolution Programs.* New York, NY: Springer.
220. Mohaghegh, S., Richardson, M. and Ameri, S. 1998. Virtual Magnetic Resonance Imaging Logs: Generation of Synthetic MRI logs from Conventional Well Logs. *SPE Eastern Regional Conference and Exhibition.* Pittsburgh, PA: SPE, SPE 51075.
221. Mohaghegh, S., Goddard, C., Popa, A., Ameri, S. and Bhuiyan, M. 2000. Reservoir Characterization through Synthetic Logs. *SPE Eastern Regional Conference and Exhibition.* Morgantown, WV: SPE, SPE 65675.
222. Rolon, L. F., Mohaghegh, S. D., Ameri, S., Gaskari, R. and McDaniel, B. 2005. Developing Synthetic Well Logs for the Upper Devonian Units in Southern Pennsylvania. *SPE Eastern Regional Conference and Exhibition.* Morgantown, WV: SPE, SPE 98013.
223. Barenblatt, G. I., Zeltov, Y. P. and Kochina, I. 1960. Basic Concepts in the Theory of Seepage of Homogeneous Liquids in Fissured Rocks. *Journal of Soviet Applied Mathematics and Mechanics,* Vol. 24, pp. 1286–1303.
224. Root, J. E. and Warren, J. P. 1963. *The Behavior of Naturally Fractured Reservoirs.* Richardson, TX: SPE, pp. 245–255.
225. Kazemi, H. 1969. *Pressure Transient Analysis of Naturally Fractured Reservoirs with Uniform Fracture Distribution. SPE Journal,* Vol. 9, pp. 451–462.
226. Rossen, R. H. 1977. *Simulation of Naturally Fractured Reservoir with Semi-Implicit Source Terms. SPE Journal,* pp. 201–210.
227. deSwaan-O, A. 1976. *Analytic Solutions for Determining Naturally Fractured Reservoir Properties by Well Testing. SPE Journal,* pp. 117–122.
228. Saidi, A. M. 1983. Simulation of Naturally Fractured Reservoirs. *Reservoir Simulation Symposium.* San Francisco, CA: SPE, SPE 12270.
229. Rubin, B. 2010. Accurate Simulation of Non-Darcy Flow in Stimulated Fracture Shale Reservoirs. *Western Regional Conference.* Anaheim, CA: SPE, SPE 132293.
230. Cipolla, C. L., Lolon, E. P., Erdle, J. C. and Rubin, B. 2010. Reservoir Modeling in Shale-Gas Reservoirs. *SPE Reservoir Evaluation and Engineering.* Richardson, TX: SPE, Vol. 13, pp. 638–653.
231. Ertekin, T., King, G. R. and Schwerer, F. C. 1986. Dynamic Gas Slippage: A Unique Dual-Mechanism Approach to the Flow of Gas in Tight Formations. *SPE Formation Evaluation.* Richardson, TX: SPE, Vol. 1, pp. 43–52.
232. Fisher, M. K., Wright, C. A., Davidson, B. M., Goodwin, A. K., Fielder, E. O., Buckler, W. S. and Steinsberger, N. P. 2002. Integrating Fracture Mapping Technologies to Optimize Stimulations in the Barnett Shale. *SPE Annual Technical Conference and Exhibition.* San Antonio, TX: SPE, SPE 77441.
233. Maxwell, S. C., Urbancic, T. I., Steinsberger, N. and Zinno, R. 2002. Microseismic Imaging of Hydraulic Fracture Complexity in the Barnett Shale. *SPE Annual Technical Conference and Exhibition.* San Antonio, TX: SPE, SPE 77440.
234. Daniels, J. L., Waters, G. A., Le Calvez, J. H., Bentley, D. and Lassek, J. T. 2007. Contacting More of the Barnett Shale through an Integration of Real-Time Microseismic Monitoring, Petrophysics, and Hydraulic Fracture Design. *SPE Annual Technical Conference and Exhibition.* Anaheim, CA: SPE, SPE 110562.

235. Kalantari-Dahaghi, A., Esmaili, S. and Mohaghegh, S. D. 2012. *Fast Track Analysis of Shale Numerical Models. Canadian Unconventional Resources Conference.* Calgary, Alberta: SPE, SPE 162699.

236. Mohaghegh, S. D. and Abdulla, F. 2014. Production Management Decision Analysis Using AI-Based Proxy Modeling of Reservoir Simulations – A Look-Back Case Study. *SPE Annual Technical Conference and Exhibition.* Amsterdam: SPE, SPE 170664.

237. Mohaghegh, S. D., Abdulla, F., Abdou, M., Gaskari, R. and Maysami, M. 2015. Smart Proxy: An Innovative Reservoir Management Tool; Case Study of a Giant Mature Oilfield in the UAE. *ADIPEC – Abu Dhabi International Petroleum Exhibition and Conference.* Abu Dhabi: s.n., SPE 177829

238. Shahkarami, A., Mohaghegh, S. D. and Hajizadeh, Y. 2015. Assisted History Matching Using Pattern Recognition Technology. *SPE Digital Energy Conference and Exhibition.* Woodlands, TX: SPE, SPE 173405.

239. Shahkarami, A., Mohaghegh, S. D., Gholami, V. and Bromhal, G. 2015. Application of Artificial Intelligence and Data Mining Techniques for Fast Track Modeling of Geologic Sequestration of Carbon Dioxide – Case Study of SACROC Unit. *SPE Digital Energy Conference and Exhibition.* Woodlands, TX: SPE, SPE 173406.

240. Amini, S., Mohaghegh, S. D., Gaskari, R. and Bromhal, G. 2014. Pattern Recognition and Data-Driven Analytics for Fast and Accurate Replication of Complex Numerical Reservoir Models at the Grid Block Level. *SPE Intelligent Energy Conference and Exhibition.* Utrecht: SPE, SPE 167897.

241. Strickland, R., Purvis, D. and Blasingame, T. 2011. Practical Aspects of Reserves Determinations for Shale Gas. *North American Unconventional Gas Conference and Exhibition.* Woodlands, TX: SPE, SPE-144357.

242. Robertson, S. 1988. *Generalized Hyperbolic Equation.* s.l.: SPE, SPE 18731.

243. Tran, N. H., Rahman, M. K. and Rahman, S. S. 2002. *A Nested Neuro-Fractal-Stochastic Technique for Modeling Naturally Fractures Reservoirs.* Melbourne: SPE, SPE 7787.

244. Akbarnejad-Nesheli, B., Valko, P. and Lee, J. 2012. Relating Fracture Network Characteristics to Shale Gas Reserve Estimation. *Americas Unconventional Resources Conference.* Pittsburgh, PA: SPE, SPE 154841.

245. Li, Y., Wei, C., Qin, G., Li, M. and Luo, K. 2013. Numerical Simulation of Hydraulically Induced Fracture Network Propagation in Shale Formation. *International Petroleum Technology Conference.* Bejing: SPE, IPTC 16981.

246. Weng, X., Kresse, O., Cohen, C., Wu, R. and Gu, H. 2011. Modeling of Hydraulic Fracture Network Propagation in a Naturally Fractured Formation. *SPE Hydraulic Fracturing Technology Conference & Exhibition.* Woodlands, TX: SPE, SPE 140253.

247. Boulis, A. S., Ilk, D. and Blasingame, T. A. 2009. A New Series of Rate Decline Relations Based on the Diagnosis of Rate-Time Data. In: *Canadian International Petroleum Conference (CIPC).* s.n.,. Calgary, Alberta. 7. Genesis.

248. Cheng, Y., Lee, W. J. and McVay, D. A. 2009. *A New Approach for Reliable Estimation of Hydraulic Fracture Properties Using Elliptical Flow Data in Tight Gas Wells.* SPE Reservoir Evaluation & Engineering, SPE 105767.

249. Mattar, L., Gault, B., Morad, K., Clarkson, C. R., Freeman, C. M., Ilk, D. and Blasingame, T. A. 2008. Production Analysis and Forecasting of Shale Gas Reservoirs: Case History-Based Approach. *SPE Shale Gas Production Conference.* Fort Worth, TX: SPE, SPE 119897.

250. Johnson, N. L., Currie, S. M., Ilk, D. and Blasingame, T. A. 2009. A Simple Methodology for Direct Estimation of Gas-in-Place and Reserves Using Rate-Time Data. *SPE Rocky Mountain Technology Conference*. Denver, CO: SPE, SPE 123298.

251. Can, B. and Kabir, C. S. 2012. *Probabilistic Performance Forecasting for Unconventional Reservoirs with Stretched-Exponential Model*. SPE *Reservoir Evaluation and Engineering Journal*. SPE 143666.

252. Ikewun, P. and Ahmadi, M. 2012. Production Optimization and Forecasting of Shale Gas Wells Using Simulation Models and Decline Curve Analysis. *SPE Western Regional Conference*. Bakersfield, CA: SPE, SPE 153914.

253. Ilk, D., Rushing, J. A. and Blasingame, T. A. 2011. Integration of Production Analysis and Rate-Time Analysis via Parametric Correlations – Theoretical Considerations and Practical Applications. *PE Hydraulic Fracturing Conference*. Woodlands, TX: SPE, SPE 140556.

254. Al-Ahmadi, H. A., Almarzooq, A. M. and Wattenbarger, R. A. 2010. Application of Linear Flow Analysis to Shale Gas Wells – Field Cases. *SPE Unconventional Gas Conference*. Pittsburgh, PA: SPE, SPE 130370.

255. Anderson, D. M., Nobakht, M., Moghadam, S. and Mattar, L. 2010. Analysis of Production Data from Fractured Shale Gas Wells. *SPE Unconventional Gas Conference*. Pittsburgh, PA: SPE, SPE 131787.

256. Nobakht, M., Mattar, L., Moghadam, S. and Anderson, D. M. 2010. Simplified yet Rigorous Forecasting of Tight/Shale Gas Production in Linear Flow. *SPE Western Regional Conference*. Anaheim, CA: SPE, SPE 133615.

257. Nobakht, M. and Mattar, L. 2012. Analyzing Production Data from Unconventional Gas Reservoirs with Linear Flow and Apparent Skin. *Journal of Canadian Petroleum Technology*, pp. 52–59.

258. Nobakht, M. and Clarkson, C. R. 2012. A New Analytical Method for Analyzing Production Data from Shale Gas Reservoirs Exhibiting Linear Flow: Constant Pressure Production. *SPE Reservoir Evaluation and Engineering Journal*, pp. 370–384.

259. Cipolla, C. L., Lolon, E. P. and Mayerhofer, M. J. 2009. Reservoir Modeling and Production Evaluation in Shale-Gas Reservoirs. *International Petroleum Technology Conference*. Doha. s.n., IPTC 13185-MS.

260. Chaudhri, M. M. 2012. Numerical Modeling of Multi-Fracture Horizontal Well for Uncertainty Analysis and History Matching: Case Studies from Oklahoma and Texas Shale Gas Wells. *SPE Western Regional Meeting*. Bakersfield, CA: SPE, SPE 153888.

261. Mayerhofer, M. J., Lolon, E. P., Warpinski, N. R., Cipolla, C. L., Walser, D. and Rightmire, C. M. February 2012. *What Is Stimulated Reservoir Volume?* SPE *Production & Operations Journal*. Vol. 25, pp. 89–98.

262. Arnsdorf, I. 2014. BloombergBusiness.com. Bloomberg.com. [Online] Bloomberg, 10 8, [Cited: January 3, 2015.] http://www.bloomberg.com/news/2014-10-07/shale-boom-tested-as-sub-90-oil-threatens-u-s-drillers.html

263. Meyer, B. R., Bazan, L. W., Jacot, R. H. and Lattibeaudiere, M. G. 2010. Optimization of Multiple Transverse Hydraulic Fractures in Horizontal Wellbores. *SPE Unconventional Gas Conference*. Pittsburgh, PA: SPE, SPE 131732.

264. Cipolla, C. L., Williams, M. J., Weng, X., Mack, M. and Maxwell, S. 2010. Hydraulic Fracture Monitoring to Reservoir Simulation: Maximizing Value. *SPE Annual Technical Conference*. Florence: SPE, SPE 133877.

265. Samandarli, O., Al-Hamdi, H. and Wattenbarger, R. A. 2011. A New Method for History Matching and Forecasting Shale Gas Reservoir Production Performance with a Dual Porosity Model. *SPE North American Unconventional Gas Conference.* Woodlands, TX: SPE, SPE 144335.

266. Bazan, L. W., Larkin, S. D., Lattibeaudiere, M. G. and Palish, T. T. 2010. Improving Production in Eagle Ford Shale with Fracture Modeling, Increased Conductivity and Optimized Stage and Cluster Spacing along the Horizontal Wellbore. *SPE Tight Gas Completion s Conference.* San Antonio, TX: SPE, SPE 138425.

267. Cipolla, C. L., Lolon, E. P., Erdle, J. C. and Rubin, B. 2010. *Reservoir Modeling in Shale-Gas Reservoirs. SPE Reservoir Evaluation and Engineering,* pp. 638–653.

268. Diaz de Souza, O. C., Sharp, A. J., Martinez, R. C., Foster, R. A., Reeves Simpson, M., Piekenbrock, E. J. and Abou-Sayed, I. 2012. Integrated Unconventional Shale Gas Modeling: A Worked Example from the Haynesville Shale, De Soto Parish, North Louisiana. *Americas Unconventional Resources Conference.* Pittsburgh, PA: SPE, SPE 154692.

269. Altman, R., Viswanathan, A., Xu, J., Ussoltsev, D., Indriati, S., Grant, D., Pena, A., Loayza, M. and Kirkham, B. 2012. Understanding the Impact of Channel Fracturing in the Eagle Ford Shale through Reservoir Simulation. *SPE Latin American and Caribbean Petroleum Engineering Conference.* Mexico City: SPE, SPE 153728.

270. Johnson, P. 1960. Evaluation of Wells for Re-Fracturing Treatments. *Spring Meeting of Southwestern District, Division of Production.* Dallas, TX: s.n, API Paper 906-5-F.

271. Mohaghegh, S. D., McVey, D., Aminian, K. and Ameri, S. 1996. *Predicting Well Stimulation Results in a Gas Storage Field in the Absence of Reservoir Data, Using Neural Networks. SPE Reservoir Engineering.* pp. 268–272.

272. McVey, D., Mohaghegh, S. and Aminian, K. 1994. Identification of Parameters Influencing the Response of Gas Storage Wells to Hydraulic Fracturing with the Aid of a Neural Network. *SPE Eastern Regional Conference and Exhibition.* Charleston, WV: s.n., SPE 29159.

273. Malik, K., Mohagegh, S. D. and Gaskari, R. 2006. An Intelligent Portfolio Management Approach to Gas Storage Field Deliverability Maintenance and Enhancement; Part One Database Development & Model Building. *SPE Eastern Regional Conference & Exhibition.* Canton, OH: SPE, SPE 104571.

274. Jacobs, T. 2015. Halliburton Reveals Refracturing Strategy. *Journal of Petroleum Technology,* pp. 40–41.

275. Siebrits, E. et al. October 2000. Refracture Reorientation Enhances Gas Production in Barnett Shale Tight Gas Wells. *SPE Annual Technical Conference and Exhibition (ATCE).* Dallas, TX: SPE, SPE 63030.

276. Hey, T. et al. 2009. *The Fourth Paradigm (Data-Intensive Scientific Discover).* s.l.: Microsoft Research.

277. Bishop, C. M. 1995. *Neural Network for Pattern Recognition.* s.l.: Clarendon Press.

278. van der Meer, L. G. H. and Egberts, P. J. P. 2008a. A General Method for Calculating Subsurface CO_2 Storage Capacity, Presented at the Offshore Technology. Conference, Houston, TX, May 5–8, OTC 19309, 887–895.

279. Plasynski, S. and Daminani, D. 2008. *Carbon Sequestration through Enhanced Oil Recovery.* Albany, OR: National Energy Technology Laboratory.

280. IPCC, Intergovernmental Panel on Climate Change. 2007. *Climate Change 2007 Mitigation.* New York, NY: Cambridge University Press.

281. IEA, International Energy Agency. 2012. *World Energy Outlook 2012*. Paris: International Energy Agency.
282. IPCC, Intergovernmental Panel on Climate Change. 2005. *Carbon Dioxide Capture and Storage*. New York, NY: Cambridge University Press.
283. Cramer, T. D. and Massey, J. A. 1972. Carbon Dioxide Injection Project SACROC Unit, Scurry County, Texas. *Annual Meeting Papers, Divison of Production*. Houston, TX: American Petroleum Institute.
284. Cameron, J. T. 1976. SACROC Carbon Dioxide Injection: A Progress Report. *Annual Meeting Papers*. Los Angelas, CA: American Petroleum Institute.
285. Marchetti, C. 1977. On Geoengineering and the CO_2 Problem. *Climatic Change*, Vol. 1, pp. 59–68.
286. Baes Jr, C. F., Beall, S. E., Lee, D. W. and Marland, G. 1980. The Collection, Disposal, and Storage of Carbon Dioxide. *Interactions of Energy and Climate*. (Eds.) Bach, W. et al. Dordrecht: D. Reidel Publishing Company, pp. 495–519.
287. Baklid, A., Korbol, R. and Owren, G. 1996. Sleipner Vest CO_2 Disposal, CO_2 Injection into a Shallow Underground Aquifer. *SPE Annual Technical Conference and Exhibition*. Denver, CO: SPE, SPE 36600.
288. Whittaker, S. 2010. *IEA GHG Weyburn-Midale CO_2 Storage & Monitoring Project*. s.l.: National Energy Technology Laboratory. Regional Carbon Sequestration Partnerships Annual Review.
289. 2015 Paris Climate Conference. [Online] December 2015. http://www.cop21paris.org
290. The Paris Agreement. [Online] April 2016. http://unfccc.int/paris_agreement/items/9485.php
291. International Energy Agency. [Online] 2015. https://www.iea.org/publications/freepublications/publication/CarbonCaptureandStorageThesolutionfor deepemissionsreductions.pdf.
292. Sengul, M. 2006. CO_2 Sequestration – A Safe Transition Technology. *SPE International Health, Safety & Environment Conference*. Abu Dhabi: SPE.
293. Kaarstad, O. 2008. Experience from Real CCS Projects – And the Way Forward. *19th World Petroleum Congress*. Madrid: World Petroleum Congress.
294. IEA, International Energy Agency. 2010. *Energy Technology Perspectives*. Paris: OECD/IEA. https://www.iea.org/publications/freepublications/publication/etp2010.pdf
295. Bureau of Economic Geology. [Online] http://www.beg.utexas.edu/coastal/
296. Ranies, M. A., Dobitz, J. K. and Wehner, S. C. 2001. A Review of the Pennsylvanian SACROC Unit. *The Permian Basin: Microns to Satellites, Looking for Oil and Gas at All Scales*. (Eds.) Viveros, J. J. and Ingram, S. M. Midland, TX: West Texas Geological Society, pp. 67–74.
297. *Crossett Devonian Field*. 1996. s.l.: Shell Oil Co. Hearing data presented to Texas Railroad Commission as reported in 1966 Secondary Recovery Application Summaries. http://www.rrc.state.tx.us/media/40380/08-0296956-pfd-shell-western-ep.pdf
298. Herzog, H., Eliasson, B. and Kaarstad, O. 2000. Capturing Greenhouse Gases. *Scientific American*, Vol. 282(2), pp. 72–79.
299. Zhang, Y. 2013. Modeling Uncertainty and Risk in Carbon Capture and Storage. *PhD Dissertation*. Pittsburgh, PA: Chemical Engineering, Carnegie Mellon University.
300. Intelligent Solutions Inc. 2013. [Online]. http://www.intelligentsolutionsinc.com/

301. McMillan, B. 2008. *Surface Dissolution: Addressing Technical Challenges of CO_2 Injection and Storage in Brine Aquifers.* Austin, TX: The University of Austin at Texas.

302. Tao, Q., Checkai, D. and Bryant, S. 2010. Permeability Estimation for Large-Scale Potential CO_2 Leakage Paths in Wells Using a Sustained-Casing-Pressure Model. *SPE International Conference on CO_2 Capture, Storage, and Utilization.* New Orleans, LA: SPE, SPE 139576.

303. Zhu, H., Lin, Y. and Zeng, D. 2013. *Oil and Gas Journal.* [Online] [Cited: 12, 2, 2013]. http://www.ogj.com/articles/print/volume-111/issue-12/drilling-production/study-addresses-scp-causes-in-co-2-injection-production-wells.html.

304. Bouquet, S. et al. September 2009. CO2CRC Otway Project, Australia: Parameters Influencing Dynamic Modeling of CO_2 Injection into a Depleted Gas Reservoir. *SPE Offshore Europe Oil & Gas Conference & Exhibition.* Aberdeen: SPE, SPE 124298.

305. Dance, T., Spencer, L. and Xu, J. 2009. *Geological Characterization of the Otway Project Pilot Site: What a Difference a Well Makes.* s.l.: Elsevier, Vol. 1, 1.

306. Bureau of Economic Geology, The University of Texas at Austin, Gulf Coast Carbon Center, 2017, http://www.beg.utexas.edu/gccc/sacroc.php

307. Myers, D.A., Stafford, P.T., Burnside, R.J., 1956. Geology of the Late Paleozoic Horseshoe Atoll in West Texas. Bureau of Economic Geology Publication 5607, p. 113.

308. Han W.Sh. Evaluation of CO_2 Trapping Mechanisms at the SACROC Northern Platform: Site of 35 Years of CO_2 Injection [Report] : PhD Dissertation. - Socorro, New Mexico : The New Mexico Institute of Mining and Technology, 2008.

309. Bachu, S. et al. 1994. Aquifer Disposal of CO_2: Hydrodynamical and Mineral Trapping. *Energy Conversion and Management* Vol. 35, pp. 269–279.

310. Preuss, M. et al. 2001: Mba1, A Novel Component of The Mitochondrial Protein Export Machinery of the Yeast Saccharomyces Cerevisiae. *Journal of Cell Biology,* Vol. 153(5), pp. 1085–1096.

311. Xu, A. et al. 2002. Novel Genes Expressed in Subsets of Chemosensory Sensilla on the Front Legs of Male Drosophila Melanogaster. *Cell and Tissue Research,* Vol. 307(3), pp. 381–392. FlyBase ID. FBrf0146927.

312. Bergenback, R. E. and Terriere, R. T. 1953, *Petrography and Petrology of Scurry Reef,* Scurry County, Texas: AAPG Bulletin, Vol. 37, pp. 1014–1029.

313. Burnside, R. J. 1959, *Geology of Part of Horseshoe Atoll in Borden and Howard Counties,* Texas: Geological Survey Professional Paper 315-B, pp. 21–35.

314. Kane, A. V. 1979, Performance Review of a Large-Scale CO_2-WAG Enhanced Recovery Project, SACROC Unit Kelly-Snyder Field. *Journal of Petroleum Technology,* Vol. 31, pp. 217–231, doi:10.2118/7091-PA.

315. Langston, M. V., Hoadley, S. F., and Young, D. N. 1988, Definitive CO_2 Flooding Response in the SACROC Unit: Tulsa, Oklahoma, USA, 16–21 April, 1988, *SPE Enhanced Oil Recovery Symposium,* doi: 10.2118/17321-MS.

316. Brnak, J., Petrich, B., and Konopczynski, M. R. 2006, Application of Smart Well technology to the SACROC CO2 EOR project: A case study: Tulsa, Oklahoma, USA, April 22–26, 2006, *SPE/DOE Symposium on Improved Oil Recovery.*

317. Schatzinger, R. A., 1988, Changes in Facies and Depositional Environments Along and Across the Trend of Horseshoe Atoll, Scurry and Kent Counties, Texas, (Ed.) Cunningham B. K., *Permian and Pennsylvanian Stratigraphy Midland Basin,* West Texas: Studies to Aid Hydrocarbon Exploration: Permian Basin Section, Society for Economic Paleontologists and Mineralogists Publication 88–28, pp. 79–95.

318. Reid, A. M. and Reid, S. A. T. 1991, *The Cogdell field study, Kent and Scurry counties, Texas: A post-mortem*, (Ed.) Candelaria, M., The Permian Basin Plays: Tomorrow's Technology Today: West Texas Geological Society Publication 91–89, pp. 39–66.
319. Tao, Q., Checkai, D., Huerta, N. J. and Bryant, S. L. 2010. Model to Predict CO_2 Leakage Rates Along a Wellbore. Florence, Italy: Society of Petroleum Engineers, 2010. ISBN: 978-1-55563-300-4.

Index

T - #0400 - 071024 - C302 - 234/156/13 - PB - 9780367734381 - Gloss Lamination